Springer
Berlin
Heidelberg
New York
Barcelona
Budapest
Hong Kong
London
Milan
Paris
Singapore
Tokyo

ADVANCES IN MATERIALS RESEARCH

The series Advances in Materials Research reports in a systematic and comprehensive way on the latest progress in basic materials sciences. It will contain both theoretically and experimentally oriented texts written by leading experts in the field. Advances in Materials Research is a continuation of the series Research Institute of Tohoku University (RITU).

H. Kitagawa T. Aihara, Jr. Y. Kawazoe (Eds.)

Mesoscopic Dynamics of Fracture

Computational Materials Design

With 151 Figures

Springer

Professor Dr. Hiroshi Kitagawa
Department of Adaptive Machine systems,
Graduate School of Engineering, Osaka University
2-1 Yamada-oka, Suita, Osaka 565-0871, Japan

Dr. Tomoyasu Aihara, Jr.
Professor Dr. Yoshiyuki Kawazoe
Institute for Materials Research, Tohoku University
2-1-1 Katahira, Aoba-ku, Sendai 980-8577, Japan

Series Chief Editor:

Professor Dr. Yoshiyuki Kawazoe

Series Editors:

Professor Dr. Masayuki Hasegawa
Professor Dr. Akihisa Inoue
Professor Dr. Norio Kobayashi
Professor Dr. Toshio Sakurai
Institute for Materials Research, Tohoku University
2-1-1 Katahira, Aoba-ku, Sendai 980-8577, Japan

Professor Dr. Luc Wille
Department of Physics, Florida Atlantic University
777 Glades Road, Boca Raton, FL 33431, USA

ISBN 978-3-642-08396-9

Library of Congress Cataloging-in-Publication Data: Mesoscopic dynamics of fracture: computational materials design / H. Kitagawa, T. Aihara, Y. Kawazoe (eds.). p. cm. – (Advances in materials research ; 1). ISBN 3-540-64291-9 (hard: alk. paper) 1. Fracture mechanics. 2. Mesoscopic phenomena (Physics). I. Kitagawa, Hiroshi, 1940- . II. Aihara, T. (Tomoyasu), 1964- . III. Kawazoe, Y. (Yoshiyuki), 1947- . IV. Series. TA409.M458 1998 620.1′126–dc21 98-3587

Printed in Germany

Cover concept: eStudio Calamar Steinen
Cover design: *design & production*, Heidelberg

Preface

This is the first volume of the new series "Advances in Materials Research" published by Springer-Verlag and edited by the Institute for Materials Research (IMR), Tohoku University. This series is a continuation of the RITU Journal which has a venerable forty year history. Although the RITU series was widely admired, e.g. Ziman's textbook mentions that an important observation was published exclusively in RITU, its circulation was limited to 1600 copies per volume, distributed mainly to libraries all over the world.

As Japan generates a wealth of experimental and theoretical materials information, IMR decided to maintain the publishing tradition but to change the method of distribution. After discussions with Springer-Verlag the present series was selected.

For the first volume the current status of computer simulation for materials research was surveyed and as topic "mesoscopic dynamics of fracture" was selected. It follows four preceeding volumes on computer simulation in the RITU series. We have organized a workshop on this developing subject and more than half of the papers in this volume have been prepared as reports of this meeting. To make this volume even more attractive to researchers and graduate students in materials science and mechanical engineering, we have asked several top researchers in this field to contribute review and complementary papers.

Beginning with this volume, we are planning to publish several volumes per year on various topics in advanced materials research. For the next volume, "Amorphous Materials XI" is expected and "Surface Science II" will follow. I would like to express my sincere appreciation to Dr. Claus Ascheron of Springer-Verlag for his great help with this new series.

Sendai
March 1997

Dr. Yoshiyuki Kawazoe
Professor
IMR, Tohoku University

Contents

Part 1

General Reviews

Challenge to Mesoscopic Dynamics of Fracture

Hiroshi Kitagawa

Department of Adaptive Machine Systems, Graduate School of Engineering, Osaka University, 2-1 Yamada-oka, Suita, Osaka 565, Japan

Summary. Computational mechanics based on a continuum model sometimes gives misleading results regarding the microscopic reality. In order to grasp the complex reality of a fracture process, this should be revised, to be able to treat the interrelationship between the micro- and macroscopic dynamics directly.

Key words. Mesoscopic dynamics, computational fracture mechanics, molecular dynamics, crack tip field, phenomenological approach

1. Introduction

Basically, fracture is a process in, or a state of, solid materials being separated due to successive failure of atomic bondings. Macroscopically, however, it is observed as a complex process which proceeds in the materials, interacting with various micro- (and meso-) scale structures. As the intrinsic strength properties of the matter which come from the most elemental structure (atomic structure) may be evaluated properly, the dynamics of the intermediate scale or mesoscale structure being involved in the process is the key to study the fracture.

In the framework of fracture mechanics, the material strength is considered to be a structure-sensitive property irrespective of individual microstructure, and there is no general idea within it to examine explicitly the complexity of fracture process.

With the fast growth of computational power as background, remarkable progress in computer simulation has been accomplished to examine the substantial nature of the materials, based on the elemental material structure and its dynamics. Computational fracture mechanics, which is being revised in response to the progress of computers, will play a great role to connect the micro- and macro-behavior of fracture properly, discussion of which, together with an introduction of some recent research results, is the object of this review.

2. Phenomenological Approaches

The scientific approach to nature is inevitably phenomenological. Two philosophical viewpoints have been known to exist historically concerning the relationship between the phenomenon and the substance, transcendental and immanent or exclusive and inclusive. In the former transcendental point of view, the phenomenon is regarded only as an appearance of the substance, and in the latter, the phenomenon represents the substance itself. If we attempt to extend the perceiving world into the imperceptible and intrinsic internal realm by logical reasoning and intend to deduce a universal knowledge through fragmentary experiences, an inclusive standpoint for the phenomenon and substance are essential in the logical framework.

Some actual examples will be examined below, in which such a somewhat conceptual statement becomes significant sharply in the mechanical research studies on fracture.

3. Methodological Problem of Computational Fracture Mechanics

Computational fracture mechanics (CFM), which has been developed so far, has a crucial methodological difficulty, because it has no means with to approach the material strength directly. Estimation of resistance of the material and analysis of mechanical circumstance to drive fracture must be treated as separate problems in CFM. What can CFM predict then concerning the material strength? Some illustrative examples to expose inconsistencies will be presented below.

3.1 What is the reality of stress field in a crack tip domain

We can obtain the details of atomic structure around a crack tip and the dynamical changes under external loading by a computer simulation using atomic modeling, e.g. the molecular dynamics (MD) method. On the assumption that the results given by such a simulation are considered to be reliable, we will evaluate what CFM predicts.

3.1.1 Atomistic Description of Crack Tip Field. Figure 1 shows an atomic arrangement near a crack tip under mode-I loading. The material model is Fe-single crystal with a bcc structure. FS potential[1] is used in the molecular dynamics (MD) analysis. The temperature is assumed to be very low and then atoms vibrate imperceptibly around their stable positions. Figure 2 indicates thedistribution of stress component σ_y in $x=0$ (plotted by "○") and y=0(by "□") planes. The stress is calculated by Born's equation[2] from the potential and the atomic configuration, which represents interatomic force per unit area. It should be noted that because of a slight asymmetry of atomic arrangement with respect to $x=0$ and y=0 planes as seen in Fig. 1, stress distributions are also asymmetric.

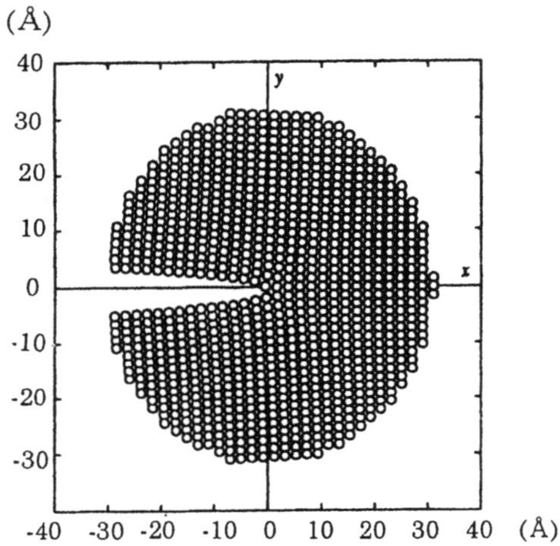

Fig. 1 Atomic configuration around a crack tip under mode-I loading (Fe-single crystal, FS potential).

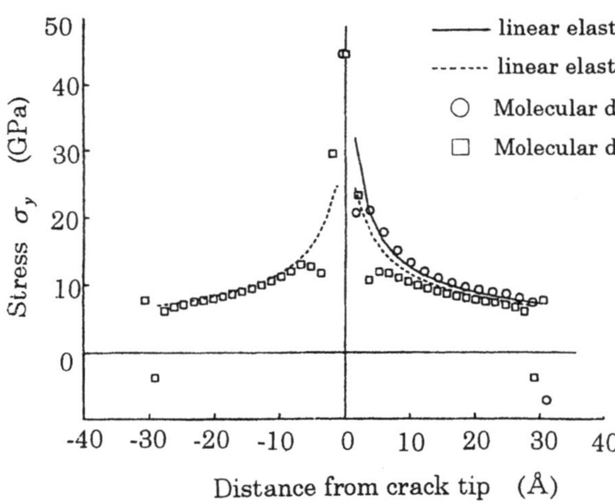

Fig. 2 Distribution of stress σ_y on $x, y=0$ planes.

Figure 3 indicates an atomic configuration and stress distribution around a mode-II crack after dislocation emission. The dislocation is subjected to a repulsive force from the crack tip field, so that it moves quickly away after the nucleation and a stress shielded zone remains between the tip and the dislocation[3].

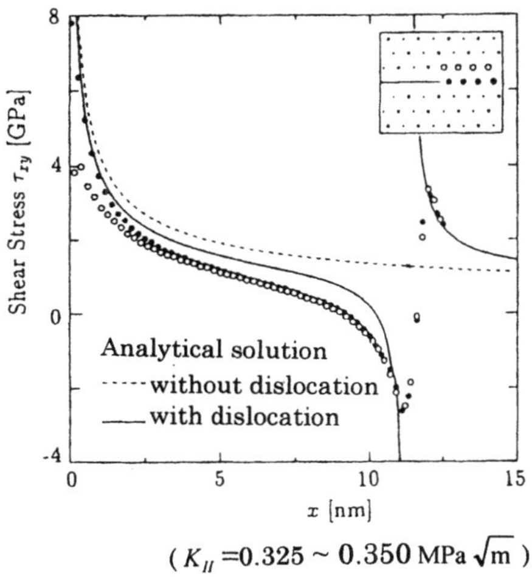

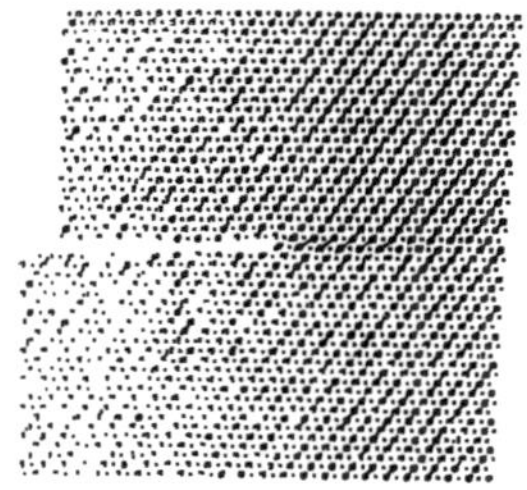

Fig. 3 Distribution of stress τ_{xy} and atomic configuration in near-tip domain under mode-II loading.

If the material has no regular, or no long-range order, structure, characteristic features of the atomic arrangement become quite different. As an example, results of MD simulation for cracks in an amorphous metal are shown in Figs. 4 and 5[4]. Due to structural inhomogeneity or randomness, stress around each atom fluctuates sharply, and a certain smoothing has been done on estimation of the local stress (strain). The amorphous is an intrinsically ductile material and local high stress induced by the external loading is relaxed easily by local atomic rearrangement. However, because there is no mechanism to propagate, or extend, the local deformation up to a macroscopic scale domain, the amorphous metal has no so-called strain hardening property and it breaks like a brittle material. Reflecting these characteristics, so-called process zone is formed clearly in the near-tip region, with a smaller dimension of almost twice the crack opening.

3.1.2 Continuum Solutions of Singular Stress Field near a Crack Tip. We can analyze the stress in the near-tip field on the basis of continuum mechanics. A continuum model, however, generally represents an averaged characteristic, so that it gives an unrealistic prediction for a local state. We first refer to crack tip fields calculated

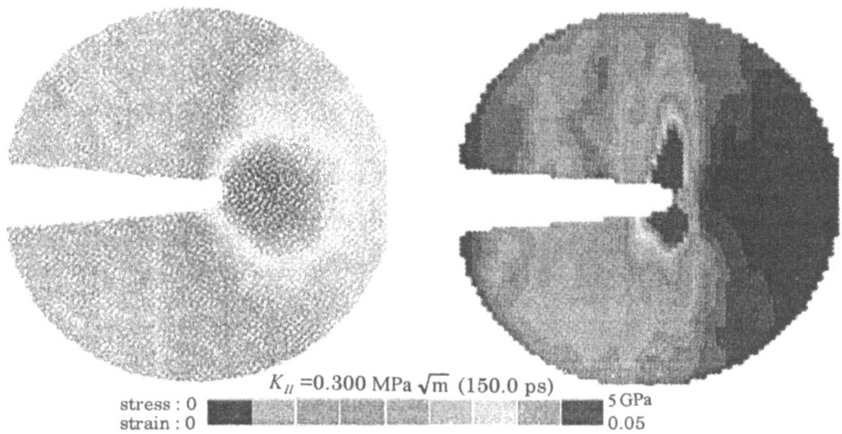

(a) Mean stress (negative hydrostatic pressure)

(b) Equivalent strain (second invariant of strain tensor)

Fig. 4 Mean stress and equivalent strain distribution around a blunted crack tip in amorphous material under mode-I loading.

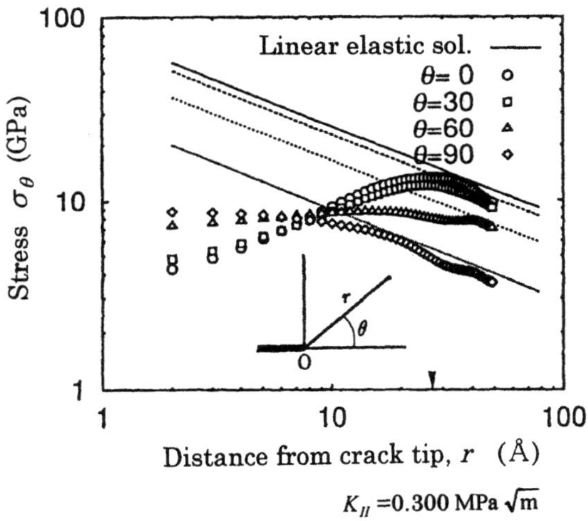

Fig. 5 Distribution of circumferential stress component along radial directions drawn from the crack tip

for three kinds of continuous media; (1) linear elastic material, (2) nonlinear elastic material whose stress--strain relation is represented by a Ramberg--Osgood type equation, and (3) Cosserat material.

In Fig. 2, the well known linear elastic solution of stress in the near-tip region is plotted, which agrees well with the results of MD-simulation in the region separated from the tip by more than 4 or 5 times the lattice constant. However, it should be noted that a remarkable discrepancy appears in the extreme neighborhood of the tip, especially in σ_y in the $y = 0$ planes.

The nonlinear elastic solution which is known as the HRR solution[5] and Cossart material solution[6] are slightly complicated and are difficult to compare as a whole with the MD--result. We focus on the ratio $\sigma_y(\delta,0)/\sigma_y(0,\delta)$, where δ is distance from the crack tip in the x or y directions, taking note that the ratio is constant for all these solutions:

$$\sigma_y(\delta,0) \Big/ \sigma_y(0,\delta)$$

$$= 0.943 \text{ for linear elastic material}$$

$$= 1.58\,(n = 3), = 1.86\,(n = 13) \text{ for nonlinear elastic material (HRR solution),}$$

$$= 0.762\ (\nu = 1/3) \text{ for Cosserat material}$$

In the HRR solution, n denotes the exponent of power-law type stress--strain relation. From the above result, the Cosserat model seems to give the most reasonable estimation even for the stress just near the tip. On the other hand, the HRR solution predicts the reverse relation concerning the ratio, although modeling of the material is more elaborate. The last example is symbolic, in which a sophisticated model gives a rather wrong result.

3.1.3 Continuum Solutions for Crack Process of Inelastic Material. In Fig. 6, the elasto-plastic solution of a mode-II crack obtained by the finite element method (FEM) based on a crystal plasticity theory[7] is indicated. The crystal lattice relative to the crack direction is the same as in Fig. 3.

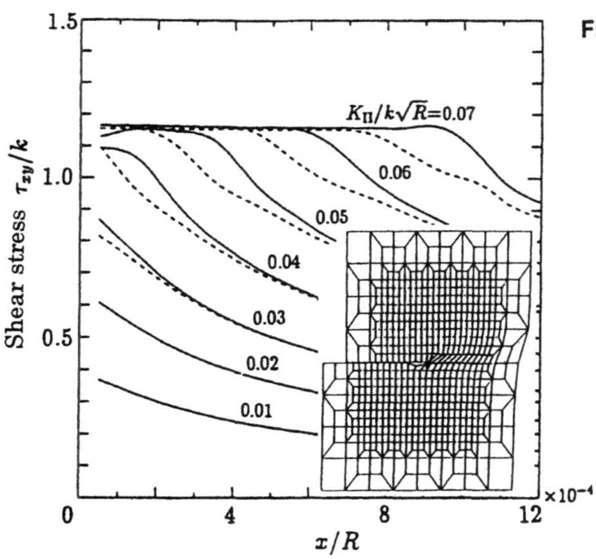

Fig. 6 Shear stress distributin on the plane in extension of crack surface (FEM analysis based on crystal plasticity theory).

Dislocations are observed to be emitted towards the extension of the crack surface in the MD-simulation, which corresponds precisely to the direction of the activated crystal-slip system in the FEM analysis. This agreement does not necessarily mean that the crystal plasticity model gives a correct prediction of the microscale structural rearrangement of a plastic material. The stress distribution obtained by FEM analysis is quite different from the one shown in Fig. 3, in which the distribution of stress component τ_{xy} is calculated by MD simulation and by the linear elasticity theory for interaction field of the crack tip and edge dislocatio[7]. Furthermore, FEM gives no information about the existence of the dislocation free zone at the crack tip.

By contrast, a continuum solution for deformation around a blunted crack based on the damage mechanics theory matches well to the MD simulation result for amorphous material. Assuming material parameters included in the constitutive equations are chosen appropriately, the FEM analysis based on the continuum damage mechanics probably gives the same stress or strain distributions as shown in Figs. 4 and 5[8].

Thus, CFM sometimes gives misleading results for the local state where fracture proceeds. The reason for this is simple, i.e., "continuum mechanics (physics) presumes nothing regarding the structure of matter"[9]. Is it improper then to estimate material strength? The answer is absolutely positive, if the following paradoxical situation is resolved.

4. Toward Computational Mesomechanics

There are few problems in mesoscopic scale phenomena which can be treated with quantum mechanics or relativistic theory. Usually, classical mechanics is the better tool to analyze the mesoscopic world. Complexity in structure and its dynamics are problems to be challenged. Remembering that both MD simulation and continuum mechanics are constructed on the basis of classical mechanics, its framework has consistency penetrating through the mesoscopic realm. Thus, it can be emphasized that the phenomenological approach, based on classical mechanics with inclusive character, has all the features necessary to connect the macroscopic and microscopic worlds.

4.1 A Paradoxical Character of Continuum Mechanics

The constitutive equations which practically represent the material properties are considered to be a problem outside the mechanical framework. In the traditional continuum mechanics, their formulation is accomplished on the basis of several principles, such as the principles of determinism, of local action, of objectivity of equipresence, of fading memory, and so on[10]. These are not mechanical, but physical or mathematical requirements, and then they are used as a conventional guide or an expedient to replace empirically obtained data in mathematical expressions. As a result, a paradoxical situation appears: as the constitutive equation is more and more sophisticated, mechanical connection between the micro-constituent and the macro-event becomes unclear. Examples of problems brought about by this have been shown earlier.

4.2 Toward Computational Mesomechanics

Usually it is said that without invoking statistical mechanics (and thermodynamics), we cannot begin to move from the fundamental theory governing the behavior of the microscopic constituents of matter to a full explanation of the macroscopic behavior in terms of microscopic constitution and its laws[11]. As continuum statistical modeling to connect discrete particle (atomic) system and the macroscopic continuous field has been well developed[12], in which no assumption resulting from outside the mechanical framework is introduced, other than the interatomic potential, it may be possible to elaborate an algorithm to treat the mesodynamics explicitly. However, the dynamics of the probability distribution function in the phase space should be treated. This necessiates treating the Liouville equation directly, which is a huge problem. The means to reduce it to a reasonable size to enable analysis by the usual computational facilities remains.

A promising approach to the mesoscale must be constructed on the basis of continuum mechanics, as expected. If we perform field analysis of the material without introducing mathematically formulated constitutive equations, the mesoscale information of the deformed state can be removed explicitly. In other words, computational mesomechanics is the key to constitute a mechanical treatment of the constitutive properties of the material.

A successful example is FEM analysis of defects in solids using a quasicontinuum model performed by Tadmor *et al.*[13]. They proposed a method for analysis of coupled atomistic/continuum deformation processes in crystals based on the use of atomistic energy funtions to describe the constitutive response and an adaptive finite element method. Although their method can treat only static and regularly ordered atomic configuration, it is significant to permit the analysis of problems requiring the simultaneous resolution of continuum and atomistic length scales in a unified manner.

Another promising approach is a linear elastic analysis including singular stress (strain) fields which represent physical constituents in the microscale (atomic-scale). If evolution dynamics of the microstructural constituents can be estimated, for example, by molecular dynamics, the properties of the mesostructure can be estimated from the distribution of the singularities. The discrete dislocation dynamics (DDD) is one such means, which will be discussed in a separate paper in this volume.

5. Concluding Remarks

Looking toward the next century, a project to develop technology for "petaflops computing" has been started. Even if such a ultra-high performance machine could be used, it would not drastically improve problems in computational estimation of material strength. Computation power is still far from that needed to analyze details of the fracture process.

Development of computer simulation, in which knowledge not to ground on mechanics does not introduce, mechanically conclusive simulation, is furthermore the key to challenge the mesoscopic dynamics of fracture.

Aknowledgement

The author gratefully acknowledges partial support from the Ministry of Education, Science and Culture of Japan under Grant-in-Aid for Co-operative Research B, Project No.09895004.

References

1. M.W. Finnis, J.E. Sinclair: Philos. Mag. A, **50-1**, pp.45~55, (1984).
2. M. Born, K. Huang: *Dynamical Theory of Crystal Lattices*, Oxford, 1954, p.129.
3. H.Kitagawa, A.Nakatani, Y. Shibutani: Mater. Sci. Eng. **A176**, 263~269, (1994).
4. These results were obtained by Dr. K. Nakatani of Osaka Prefectural

University, details of which will appear in this volume.
5. J.W. Hutchinson: J.Mech.Phys.Solids, **16**, 13~31,(1968). J.R. Rice, G.F. Rosengren: J.Mech.Phys.Solids, **16**, 1~12, (1968).
6. R. Sternberg, R. Muki: Int.J.Solid and Struct., **3**, 63, (1967).
7. H. Kitagawa, A. Nakatani, Y. Shibutani,: Acta Mech. Solida Sinica, **8**, 53~57, (1995).
8. Detailed discussion on this will be found in the paper of K. Nakatani in this volume.
9. C. Truesdell, W. Noll: *The Non-Linear Field Theories of Mechanics*, 2nd ed., Springer-Verlag, 1992, p.5.
10. A.C. Ringed: *Mechanics of Continua*, John Wiley & Sons, Inc, 1967, p.142.
11. L. Sklar: *Physics and Chance, Philosophical Issues in the Foundations of Statistical Mechanics*, Cambridge University Press, 1993.
12. For example, J.H. Irving, J.G. Kirkwood: J. Chem.Phys.,**18-6**, 817~829, (1950).
13. E.B. Tadomor, M. Ortiz, R.Phillips: Philos. Mag. A, **73-6**, 1529~1563, (1996).

Large-Scale Atomistic Simulations of Fracture

Luc T. Wille

Department of Physics, Florida Atlantic University, 777 Glades Road, Boca Raton, FL 33431, USA

Summary. This paper discusses the use of realistic, microscopic, molecular dynamics simulations of crack initiation and propagation in metallic systems. Some of the outstanding problems are outlined and the atomistic approach is compared and contrasted to continuum elasticity theory. Representative snapshots of some of the latest work on two- and three-dimensional systems are displayed and discussed. Attention is also given to the problems posed by alloys and grain boundaries. Finally, a critique of the current state-of-the-art is given and some speculations are offered about future trends.

Key Words. Fracture, molecular dynamics simulations, metals, metallic alloys

1. Introduction

The sudden catastrophic failure of a material due to crack formation and propagation is a process with evident economic impact. Tremendous financial savings could be obtained and tragic loss of life could be avoided if these fracture phenomena were better understood. Consequently, fracture mechanics has become an important field in mechanical engineering and materials science [1-4] with a great tradition behind it. Physicists, however, for a long time shied away from the subject, only in the last two decades gradually turning their attention to it, with the last five years witnessing a veritable boom in insights and productivity [5-7]. Many seminal papers in this area are now published in the physics literature, as befits a subject that lies at the interface of solid state physics, non-equilibrium thermodynamics, and quantum mechanics. The types of investigations range from continuum elasticity theory [1], to model studies (using bond-breaking network models, diffusion limited aggregation, etc.) [5, 7], and realistic large-scale simulations (the topic of this review).

Many of the early theoretical insights came from analytical work on continuum models of elasticity. Later, continuum models were further studied by computational means, employing such techniques as the finite element method. These approaches, which totally neglect the underlying atomistic lattice-structure of the material, have yielded many useful findings and continue to be of great use, but are ultimately limited. Not only is such a philosophy fundamentally flawed in that the cohesion of a material is due to interatomic forces, quantum mechanical in origin, acting between massive particles, but it is also restricted in the range of phenomena it can treat. For example, crystal imperfections such as impurities, dislocations, grain boundaries, etc. are very hard to include realistically in a continuum approach. Moreover, interatomic potentials are nonlinear and anisotropic, while continuum elasticity usually assumes linear and isotropic behavior. A discrete microscopic description allows one to define local temperatures, potential energies, and stress distributions, quantities that have a precise physical meaning and that are key to understanding fracture. Continuum elasticity is also beset with problems involving singularities, especially near the crack tip, while the discreteness at the atomic level provides an automatic cut-off for divergent expressions. Thus, there are many reasons, both theoretical and practical, to turn to an atomistic description of fracture.

Atomistic studies have the advantage that they deal with the precise structure (including defects) and composition of the material one wants to study. They allow one to track individual particles, as well as analyze collective motions, in order to gain a true microscopic picture of crack initiation and growth. At least in principle, one can use as accurate a description as is needed of the interatomic forces, which are ultimately quantum mechanical in nature. There are shortcomings to the atomistic approach as well: length and time scales that can be simulated with current computers are not compatible with the macroscopic phenomenon of fracture. Also, the amount of data produced may be so overwhelming that trends and features become hard to discern. Since the continuum methodology is widely used, notably in industry, and has been very useful, one of the aims of the atomistic descriptions must be to make a connection between the micro- and macro-scale and to provide microscopically derived values for the parameters that enter the continuum models.

At the simplest level, in order to understand fracture, one can distinguish two classes of materials: brittle and ductile. Materials in the first class tend to shatter on impact because cracks that are stressed will propagate unabatedly and cut through atomic bonds in a razor-like fashion. Ductile materials, on the other hand, deform on impact as cracks blunt and only propagate with great difficulty. Thus, brittle materials tend to break in the elastic regime, while ductile systems break in the plastic regime. From an energetic point of view these two categories differ in the relative magnitude of the energies associated with bond breaking (cleavage fracture) and plastic shear. Rice and Thomson [8] have proposed a simple model in which the shear is assumed to be associated with formation of a dislocation emerging at the crack tip under loading. Such dislocations would blunt the tip indicating a ductile material. Brittleness and ductility are not just materials properties but also depend on pressure and temperature, which is not unexpected within the Rice-Thomson model since the relative energies involved depend on these external parameters. Determining when a given material will be brittle or ductile is one of the outstanding problems in fracture mechanics. A number of improvements to the Rice-Thomson model have recently been proposed [9]. A famous criterion to decide whether a crack will propagate in a brittle material under loading was developed by Griffith in 1920 [1, 10] and also considers an energy balance, this time between the potential energy stored by the application of an external force and the surface energy created when bonds are broken. The Griffith criterion states, in essence, that cracks will grow if the decrease in potential energy exceeds the cost in surface energy. According to continuum mechanics the dynamic crack should remain smooth and accelerate until it reaches a limiting speed, known as the Rayleigh speed (v_R). Experimentally, one often observes that although propagation is initially one-dimensional, the cracks start to zig-zag at speeds well below v_R leading to a rough crack surface, and that their speed saturates at a fraction of v_R. Recently, some work has been done to explain these observations in terms of dynamical instabilities within the context of continuum elasticity theory [11], but it is generally felt that an atomistic explanation is needed [12, 13].

2. Atomistic Simulations: Methodology

2.1 Molecular Dynamics

The most fundamental approach to the atomistic modeling of fracture would involve the solution of the time-dependent Schrödinger (or Dirac) equation for a sufficiently large system of atoms and would include both electronic and atomic excitations. Such a line of action is clearly out of the question, both on theoretical grounds and for practical reasons such as computer resources. A less ambitious level of description might use a simplified quantum mechanical Hamiltonian, such as that provided by the tight-binding method, but even such schemes are not yet feasible. Because of the size of the systems necessary to get accurate results, one is forced to adopt a classical approach in which the atoms are assumed to be interacting via a fairly simple potential energy function and electronic degrees of freedom are not explicitly considered. The motions of the atoms may then be followed by means of molecular dynamics (MD) simulations in which the equations of motion for the system are numerically integrated.

The MD technique is a very natural one and has been used in a wide variety of situations and for a wide range of systems [14-16]. In a sense, this dates back to the days of Newton when astronomers would manually calculate the paths of planets around the sun. As an illustration, Feynman works out a pedagogical example in his Lectures on Physics [17]. In a more restricted manner, the method first came to the forefront of solid state physics exactly 40 years ago after the publication of the seminal paper of Alder and Wainwright [18]. In order to derive the relevant equations one may start from Hamilton's equations or more simply from Newton's second law, in which the equation of motion for the i-th atom is given by:

$$m_i \frac{d^2 \boldsymbol{r}_i}{dt^2} = \boldsymbol{F}_i \tag{1}$$

where m_i is the atomic mass of the i-th atom, $\boldsymbol{r}_i$ is its coordinate, and $\boldsymbol{F}_i$ is the total force acting on this atom (implicitly this involves a summation over neighboring atoms). Other dynamic equations may, of course, be considered, for example by including Langevin terms or damping forces. To proceed, one needs to discretize the right-hand side and to arrange the equation in such a way that the position and velocity at time $t + \Delta t$, can be calculated from the positions, velocities, and forces at time t. A number of schemes are available to accomplish this, including the Verlet (or Stormer) method and the Nordsieck algorithm. Both of these are numerically stable as long as the time step Δt is not too large and the choice of technique is largely a matter of preference. Methods for choosing optimal timesteps are also well documented [14, 16]. A useful test on an MD code's performance, in cases described by (1), is to check that the system's total energy is conserved (within small fluctuations). One typically starts the system off with atoms at or near their lattice positions and with velocities chosen from some distribution meant to reflect the temperature of interest. Usually, one needs to relax the system for a number of time steps before the simulation as such starts. Overall temperature may be determined from the kinetic energy via the equipartition theorem. If necessary, velocities may be rescaled to change the temperature. In order to quench a system to very low temperatures, it may be advantageous to look for potential energy crossings through a minimum and to set all velocities equal at such times. This will tend to quench the system very rapidly to close to absolute zero.

For fracture simulations a number of practical problems arise due to the sizes of the systems that need to be considered and the duration of the simulations which tend to lead to very time-consuming and computer-intensive calculations. Typically, a starter notch is introduced somewhere in the material, a strain is imposed on the system, and the dynamic response is studied. Practical problems arise due to the choice of boundary conditions. Periodic boundary conditions, which are often put to good use in MD simulations to eliminate or reduce finite-size effects, cannot be used here, except perhaps in one of the directions. Usually, free boundary conditions are used in all directions. Moreover, fracture is associated with energy release due to bond-breaking, which leads to shock waves being emitted from the vicinity of the crack tip. Such shock waves may reflect when they come to the boundary and lead to a destructive interference with the outgoing shock wave. In some cases, closure of the crack due to the reflected shock wave may be observed. Thus it is necessary to include some type of mechanism to allow for energy absorption at the boundary. Precisely how to do this in an optimal fashion remains an open problem.

2.2 Potential Construction and Choice

The construction of an interatomic potential that accurately models a given material is more of an art than a science in that there is no rigorous procedure to go from the quantum domain (electron wave functions) to the classical regime (interatomic interactions) [19]. For metals, the most widely used potentials are those constructed by means of the embedded atom method (EAM) [20]. The EAM methodology is just one of a wide variety of approaches based on an effective medium description. The key quantity is the energy needed to embed an atom into the material as described by the local electron density. Thus, it includes electronic structure information in some averaged fashion, although the result is a classical potential function. It does, however, go beyond simple pair potentials in that it contains many-body contributions. The general form of the total energy in the EAM method is given by:

$$E_{\text{tot}} = \sum_i F_i\Big(\sum_{j\neq i} f_j(r_{ij})\Big) + \frac{1}{2}\sum_{ij}\phi_{ij}(r_{ij}) \tag{2}$$

where i and j denote atoms, F_i is the embedding energy for the i-th atom and its argument is the total electron density due to the other atoms in the system. In principle, this involves a sum over all atoms, but in practice, in a metal, this contribution is short-ranged due to screening effects. The ϕ_{ij} is also short-ranged and takes into account the core-core repulsion. It is often assumed to be of the form of a truncated Lennard-Jones potential. Specific prescriptions for determining these functions for a specific system can be found elsewhere [20]. We also note that this and the other potentials described here may be brought to a dimensionless form, by choosing appropriate length and time scales. EAM potentials have been used in several fracture studies [21-25], but their evaluation is rather time consuming. In view of the large number of atoms needed in accurate studies and the duration of the simulations, it is essential to use as simple a potential function as possible, whilst still obtaining meaningful results. Such a description is provided by the simple truncated Lennard-Jones (LJ) potential, in which the potential between two atoms a distance r apart is taken to be of the form:

$$\varphi_{\mathrm{LJ}}(r) = 4\varepsilon \left[\left(\frac{\sigma}{r}\right)^{12} - \left(\frac{\sigma}{r}\right)^{6} \right] \tag{3}$$

where ε is the energy well depth and σ the point at which the potential goes through zero. Typically, this potential is truncated by a smooth cut-off [22]. The cut-off is a spline function which ensures a smooth approach to zero at a finite distance and keeps the potential short-ranged so that the force calculation only runs over the atoms in a neighboring zone rather than over the entire system. Such potential functions are now commonly used in many of the largest MD simulations of fracture [22, 26-30]. It is generally argued that fracture is such a violent process that the details of the potential (i.e. EAM, instead of LJ or some other form) do not matter, as long as one obtains an accurate description of the well depth and position of the minimum in the potential. To some extent this may be wishful thinking, but it seems clear that in order to get meaningful results in a reasonable time with current computers, it is more important to increase the size of the system and simplify the potential, rather than take a more complex potential but for a smaller system. It is also worth pointing out that a change in the shape of the cut-off tail allows one to change the potential from one appropriate for a brittle material to one for a ductile material [22], at least in two dimensions. In three dimensions the ability of the LJ potential to represent a brittle material is not so clear.

Instead of a LJ pair potential, other authors [31-32] have proposed to use a Morse pair potential, which has the general form:

$$\phi(r) = \exp\left(-2\alpha(r\text{-}1)\right) - 2\exp\left(-\alpha(r\text{-}1)\right) \tag{4}$$

in which the nearest-neighbor distance is taken as the unit of length and the well depth is taken as the unit of energy. The parameter α is material dependent, a value of 3.5 being appropriate for Cu, while $\alpha = 6$ leads to a potential that is virtually identical to the LJ potential.

Other potentials have also occasionally been used in the literature. For example, the earliest MD simulation of fracture [33] used a potential that consisted of two quadratic curves with equal but opposite spring constants joined at an inflection point and truncated at some finite distance. This is sometimes referred to as the Hoover-Ashurst-Moran (HAM) potential. It has the advantage of leading to very simple expressions for the forces, which are therefore fast to evaluate by computer, but has fallen somewhat out of favor because of the exclusion of anharmonic effects. Other authors [21] have used the Johnson-potential as a model for α-Fe [34]. Both this and the HAM potential correspond to brittle materials. Although most of this review focuses on metallic systems, appropriate potentials for other materials have also been derived and have been used in fracture studies [35, 36]. Finally, we note that the problem of a random binary alloy (A_cB_{1-c}) poses its own complications in that now three potential functions are needed: one each for the A-A, B-B, and A-B interaction. The EAM approach has specific prescriptions for treating alloys, but as a simple prescription for the LJ potential one can take the A-A and B-B interactions to be those appropriate for the pure systems (with ε and σ determined from the cohesive energy and equilibrium bond length, respectively), while the A-B interaction can be derived from the ordered AB compound if such a system exists, or else (for a phase separating alloy) by simply interpolating the bond lengths and cohesive energies for the pure elements.

2.3 Parallel Algorithms

Simulations on the scale needed to realistically model crack propagation by necessity involve systems with several hundreds of thousands of atoms, the current record being a 400 million atom simulation. Clearly, such massive number crunching can only be undertaken on the fastest computer systems, virtually all of which are presently based on a parallel programming paradigm. Fortunately, MD simulations are well suited for parallel implementation and the various strategies for different types of architectures and potentials are well documented [37-41]. For fracture studies of metals the currently used potentials are short-ranged. This is due to the physics of the bonding, but it has the advantage of keeping the force calculation an $O(N)$, rather than $O(N^2)$ problem (N is the number of atoms in the system being simulated). The latter scaling would be prohibitive for the type of system sizes needed for accurate fracture studies.

Parallel computer architectures may be divided in two broad classes: Single-Instruction Multiple-Data (SIMD) and Multiple-Instruction Multiple-Data (MIMD). The former group of computers typically consists of several thousands of processor elements arranged in some simple geometric structure (grid or hypercube) all executing the same computer instruction on their local data elements. Examples of SIMD computers include the Connection Machine CM-2 and the MasPar MP-1 and MP-2 machines. MIMD computers on the other hand, such as the IBM SP2, Cray T3D or T3E, or farms of workstations communicating through software directives such as PVM or MPI, allow each processor to execute its own commands independently of all the others. Thus, different processors may well be performing different operations. Synchronization and load balancing become important issues in MIMD computers, although they do provide greater flexibility than SIMD machines. Efficient parallel computing is a constant struggle between communication and computation. Generally speaking, one should try to minimize the former and maximize the latter.

Two natural strategies suggest themselves for parallelizing MD simulations: atomic decomposition and geometric decomposition. In atomic decomposition, all processors get the same number of atoms assigned to them (assuming that N is an integer multiple of the number of processors), but there is no necessary geometric relation between atoms assigned to the same processor, i.e. physically nearby atoms may well end up on processors that are far apart in the interconnection network. The atoms assigned to a given processor remain so during the entire duration of the simulation. This strategy tends to lead to severe communication problems, which generally more than offsets the gains obtained by guaranteeing load balancing during the simulation. In geometric decomposition, each processor has a region of space assigned to it and consequently performs the computations for all the atoms residing in that region. As atoms move about in the course of the simulation, they are shifted from processor to processor as needed. Thus, the communication needs are always between nearby processors (in view of the short range of the potential), which maximizes processor usage. The potential load imbalance is usually not so severe in fracture problems in which the vast majority of the atoms never stray very far from their initial positions. Generally, geometric decomposition is the preferred route for fracture MD. It may be combined with other standard techniques to speed up MD simulations, such as Verlet neighbor lists [16], to lead to an efficient portable scheme.

2.4 Implementation Details

Details of the implementation depend on the computer used as well as the system studied and may be found in some of the references quoted. However, two practical aspects of MD simulations of fracture have recently received attention in the literature and are worth discussing in detail. The first issue relates to the use of random (or rather, pseudorandom) numbers used to initialize the system. Although MD simulations are completely deterministic (unless noise-terms are used, which is not the case in fracture simulations), the system is typically initialized with random initial velocities and may also have initial positions that are randomly perturbed from a perfect lattice position. Most of the high quality sequential random number generators are recursive in nature, i.e. starting from an initial value ('seed') one generates a sequence of random numbers, the $(n+1)$-th one being given by applying some fairly simple relation (often a linear congruence) to the n-th one. The natural way to port this to a parallel computer is to have each node calculate its own random numbers and to give the various generators a 'staggered' start [42]. While adequate for computer systems that consist of just a few processors, for larger parallel machines this procedure may incur considerable overhead and therefore Holian *et al.* [43] have suggested an alternative method based on shuffled, nested, Weyl sequences (SNWS). Using a sophisticated test that is able to detect correlations between non-neighboring terms in a random number sequence it was shown that the SNWS leads to minimal correlations, unlike less sophisticated schemes. These authors also display an interesting case in which a random number generator with correlations leads to unphysical results. Clearly, some care needs to be taken in MD simulations of fracture to ensure that the random number generator used does not produce any artifacts, which may well be long-lived.

A second practical aspect of MD simulations is the need to tackle the overwhelming amount of data that is being produced, a phenomenon aptly termed 'data glut' by Beazly and Lomdahl [44]. A simple estimate shows that a simulation with 100 million atoms (not at all unrealistic by today's standards) leads to a coordinate list (to be used, e.g., to generate 'snapshots') of about 2 Gbytes. Moreover, to have a clear picture of the dynamics involved one may well want to display a sequence of several hundred of such snapshots. This poses extreme demands not only on the computational resources (both in terms of data storage as in processing terms), but also on the ability of the human eye to detect the salient features. Beazly and Lomdahl [44] argue that the processing of the data can most easily be accomplished by the use of scripting languages. In order to cut back on memory requirements, it is customary to only store the data for atoms that are of interest, for example, those that are undergoing significant displacements from their relative lattice positions as measured by their total potential energy (useful in analyzing dislocation loops, etc.). The large amount of data generated also means that visualization and animation are the tools of choice for the human user to get a feeling for the dynamics of the crack propagation. Having only those atoms that are of interest being displayed makes it easier to track the system. However, as a precaution in the case of a computer crash or to analyze alternative scenarios, it may be useful to store the positions and velocities for the entire system from time to time in case one wants to re-start a run at a certain point in the system's evolution. This avoids having to commence the entire simulation run all over again.

For further details about the implementation of the MD method to fracture problems the reader is referred to the papers quoted in the list of references. For those interested in parallel computing aspects, attention is also drawn to the paper by Plimpton [39] in which detailed timing comparisons for a variety of computers and methods are given.

3. Results and Discussion

3.1 Two-Dimensional Systems

The earliest MD study of fracture focused on a two-dimensional triangular system [33]. Because of the computational demands due to the need for a large enough simulation size the two-dimensional problem has continued to receive a good deal of attention. Much of this work considers Mode-I loading in which an initial sharp triangular opening is introduced on the edge of the material and a constant outward strain rate is applied perpendicular to this notch. Fig. 1 shows a series of snapshots indicating the time evolution of such a system [29, 30]. This particular figure shows results obtained using a truncated LJ-potential to model a disordered $Cu_{0.28}Ni_{0.72}$ alloy (known as Monel). The simulation box contains 224727 atoms initially placed in a triangular lattice and cooled to 8 K. The time unit in these simulations was $t_0 = 1.195\times10^{-13}$ s. The integration step Δt equaled 0.015 t_0, which was established to lead to a stable algorithm. The simulations were performed on a MasPar MP-1 parallel computer. A constant strain rate of $10^{-4}/t_0$ was applied until crack growth occurred. The expansion was then stopped but adjustments at the boundary were made to absorb outgoing shock waves [24]. This snapshot is very similar to others obtained for different systems [24, 28]. One notes the smooth motion of the crack tip at early times, followed by a blunting and a tendency to zig-zag at later times. To quantify these observations it is useful to analyze the position and velocity of the crack tip as a function of simulation time. These data are shown in Fig. 2. In order to obtain an objective measure for the crack tip location, some interesting problems occur. For example, as can be seen in Fig. 1, as the crack proceeds, assignment of its position becomes ambiguous. Voids may be created ahead of the tip, or bridges may form closing off regions where the crack tip has already passed. In order to avoid such uncertainties, an averaging scheme was used to determine the position of the crack tip [29]. The derivative of this curve (not shown) would give the crack tip velocity and is also smooth, in contrast to many papers in which this quantity appears very noisy. Also shown in Fig. 2 is the magnitude of the x-component of the velocity of the most energetic atom in the vicinity of the crack tip. Large peaks in this quantity correspond to bonds breaking and atoms 'snapping' back. As the crystal is expanded, the nominal crack tip moves gently forward until near time step 1750 at which there is a sudden change in slope in the crack position, which corresponds to the first bond breaking (signaled by a peak in the maximum velocity curve). This occurs at an expansion of about 2.6 %. In comparison, for the elemental materials Abraham *et al.* [27] report 3 % as the critical strain, while Holian and Ravelo [24] find 2 %. After these initial events, the slope returns to that for expansion, until at time step 2020 several bonds start breaking and the crack velocity increases dramatically. The expansion at this point was 3%. From this point onward the crack accelerates and propagates through the sample as shown in Fig. 1. The speed of propagation levels off near 0.6 of the Rayleigh speed, in agreement with the observations of other simulations [27, 28]. The crack initially moves linearly, but as its speed becomes more erratic it starts to emit dislocations near the tip, leading to more of a zig-zag motion (visible in Fig. 1). However, the crack motion is not arrested and thus the material still fractures

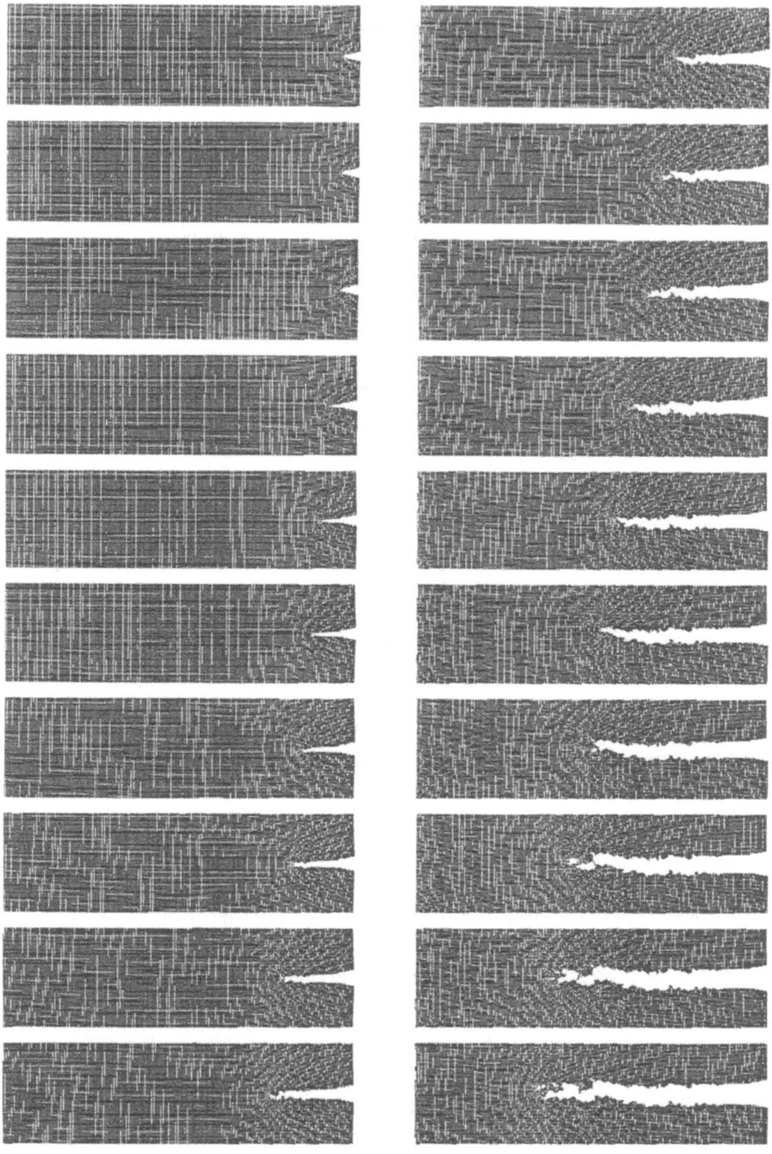

Fig. 1. Central portion of simulation cell indicating the time evolution of an initial starter notch in a two-dimensional system with constant applied strain rate. The material modeled is a Cu-Ni alloy at an initial temperature of 8 K. The first snapshot is at t = 180 ps and the time between snapshots is 30 ps.

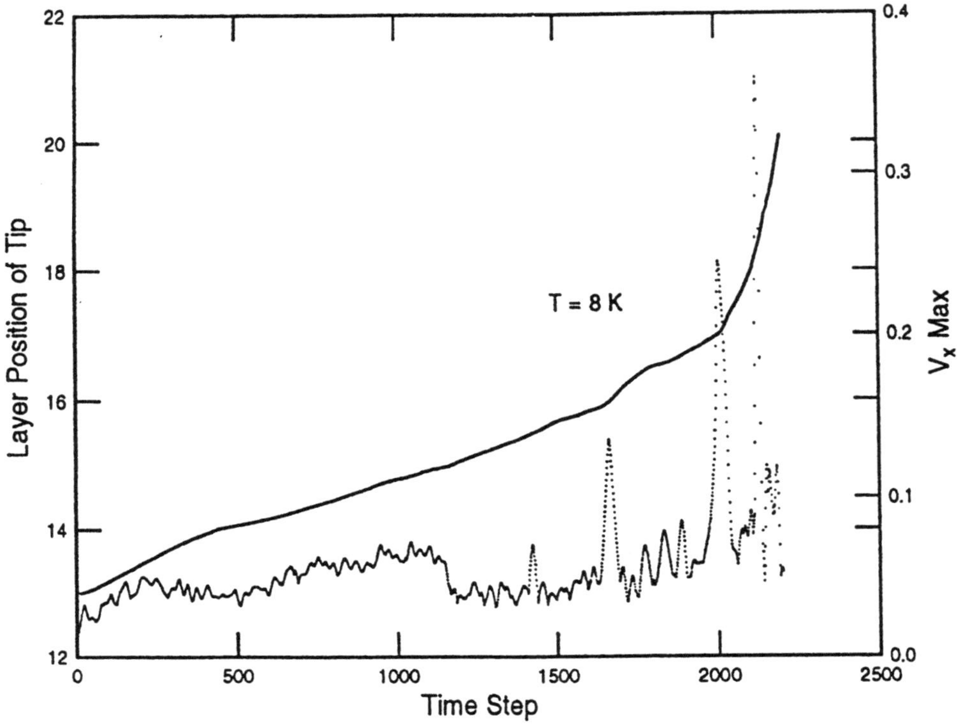

Fig. 2. Crack tip position and maximum particle velocity as a function of time (in units of the integration time step). These data correspond to the same parameters as in Fig. 1.

in a brittle fashion. The crossover in propagation mode is the dynamic instability anticipated by more sophisticated continuum mechanics treatments and now found by several MD simulations. It was also observed in the Cu-Ni case that the time at which the first bond breaks moves to lower values as the temperature of the crystal is increased, i.e. the critical strain rate is less at higher temperatures [30]. This is in contrast to a pure LJ-system in which the critical strain appears to be temperature-independent. An analysis of the particles' speed near the crack tip demonstrates clearly outgoing shock waves, generated by the bonds breaking. These disturbances move nearly at the speed of sound for the material, in a way comparable to that for the pure element [24].

3.2 Three-Dimensional Systems

As computational power has increased, the last few years have seen an increase in the number of three-dimensional simulations that are being reported. Nevertheless, these simulations are at the cutting edge of what is possible in computational materials science, making huge demands in terms of computer memory and computation time. The generalization to three dimensions is quite important as a number of artifacts in the two-dimensional case will disappear. Typical results are shown in Fig. 3, which is reproduced from the work of Abraham [28]. This simulation has 100509696 atoms in it, consisting of a three-dimensional slab with 336 atoms in both the x- and y-direction (horizontal and vertical in figure) and 896 atoms in the z-direction (pointing towards the reader). A notch is introduced midway along the x-direction, oriented along the positive y-direction with a length of 120 atoms and extending all along the z-direction. Periodic boundary conditions were imposed between the xy-surfaces at $z = 0$ and $z = L_z$. These simulations were run on an IBM RS/6000 SP parallel machine with 306 nodes and employ the LJ potential. The slab was quenched to absolute zero temperature and a constant strain rate was imposed in the negative and positive x-direction. The solid was observed to fail at an expansion of 1.5 %. The top two snapshots in Fig. 3 show the crack propagating in a brittle manner (only atoms perturbed from their nominal lattice positions are displayed here). There is a small area right at the crack tip where bonds are breaking, called the 'process zone', but otherwise the region away from the cleavage surface is unaffected. Near time step 43 (left, middle snapshot in Fig. 3) a transformation occurs in the process zone as dislocations are being emitted. As time progresses (remaining snapshots in Fig. 3) more and more dislocations appear and dissipate elastic potential energy, but the crack has come to a complete stop indicative of a ductile crystal. These dislocations move at approximately one quarter of the longitudinal speed of sound and terminate on the top surface as steps. Prior to dislocation emission the crack surface has started to roughen (not visible in Fig. 3), similar to the dynamic instability observed in the two-dimensional case. However, in the present case, this instability is associated with the brittle-ductile transition, in sharp contrast to the two-dimensional case where such a transition was not observed. This highlights the important difference between two and three dimensions as far as fracture is concerned.

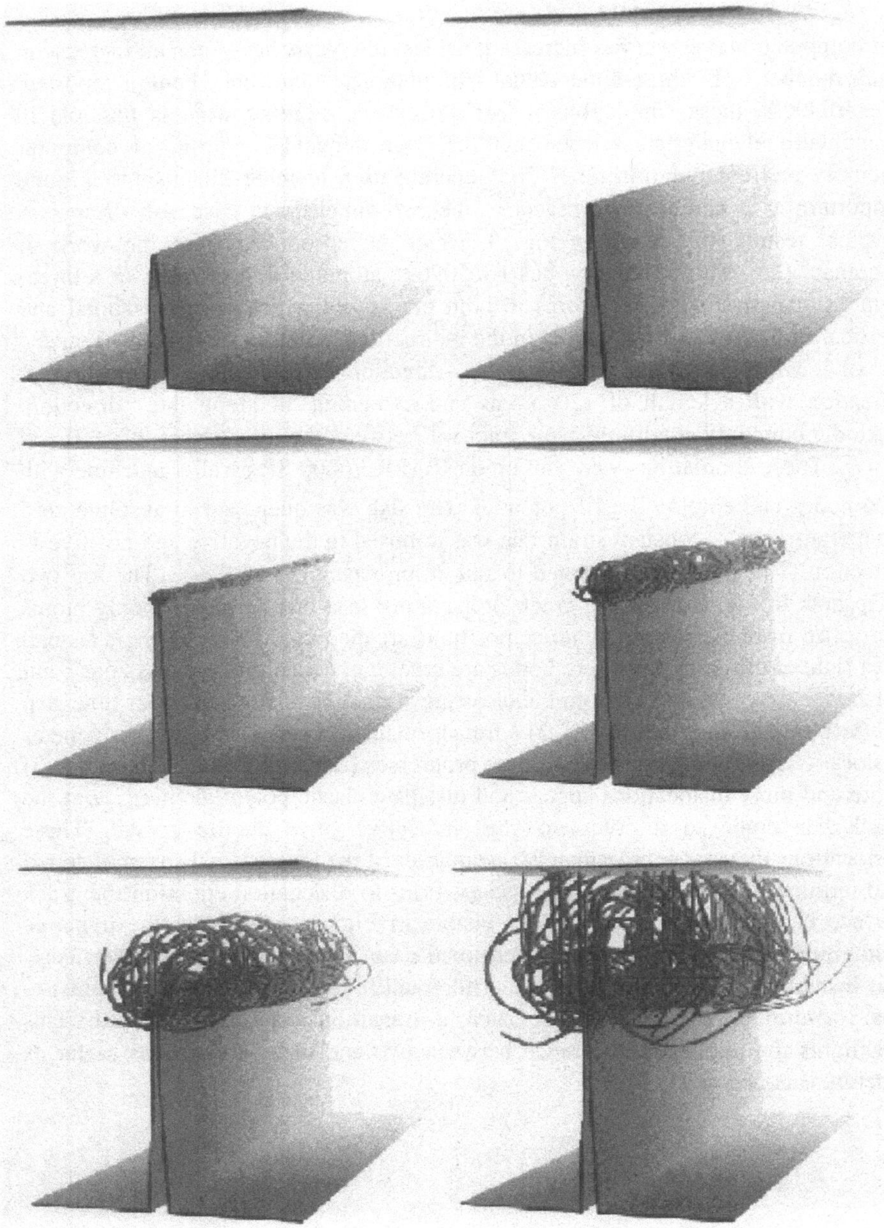

Fig. 3. Snapshots displaying crack and associated dislocations in a three-dimensional LJ-solid (100 million atoms in a slab) at times 25, 34, 43, 52, 68, and 90 in reduced units. Only atoms whose potential energy deviates by more than 3% of that in the bulk are displayed. The top and bottom surfaces of the slab are indicated, but the other exterior surfaces are suppressed. (After [28].)

3.3 Effects of Grain Boundaries

Real materials at finite temperature contain numerous defects. Perhaps the most important consideration as far as fracture is concerned is the presence of grain boundaries. Technologically it is very important if cracks encountering a grain boundary (GB) continue in a transgranular fashion or whether they propagate intergranularly. An excellent overview of this problem is given by Yip and Wolf [45]. Fracture near GBs has not received the same amount of attention from large-scale MD simulations as the ideal crystal (with initial crack). Wagner and Holian [23] in a two-dimensional MD study of spallation looked at the effects of a grain incorporated in a matrix. They found that at intermediate impact velocities, voids would open up along the GB but the inside of the grain would be largely unaffected, while at high impact velocities voids opened up throughout the sample.

Fig. 4 shows the results of a simulation for the same Cu-Ni alloy studied in Figs. 1 and 2, but this time with a Σ-5 GB present [30]. The orientation of the sample is such that the GB is perpendicular to the initial crack direction. Quenching of such a system is considerably more complicated as the lattice mismatch when the two grains are put together can be quite large, leading to large amounts of excess energy that need to be eliminated. An additional complication for an alloy is that segregation of one of the species will preferentially occur to the GB [46]. In the case of Cu-Ni, Cu is known to preferentially occupy sites near the interface. This effect was incorporated by estimating the concentrations away from the GB plane by interpolating the EAM calculations of Foiles [47] for the Cu-Ni system with various GBs. As can be seen in Fig. 4, the crack initially propagates in much the same manner as in the single crystal alloy (Fig. 1), but as the crack tip approaches the boundary it slows down and starts to meander due to the interaction with the GB. The crack blunts and essentially comes to a halt at the GB, but some voids are created along the GB. Clearly it would be interesting to study other types of boundaries as well as other alloy systems and compositions. Moreover, the effects of the third dimension could be quite important in the presence of GBs. This seems to be a fertile area for future research.

4. Conclusions

It should be clear from this brief review that computational materials science has only barely started to have an impact on the important problem of crack initiation

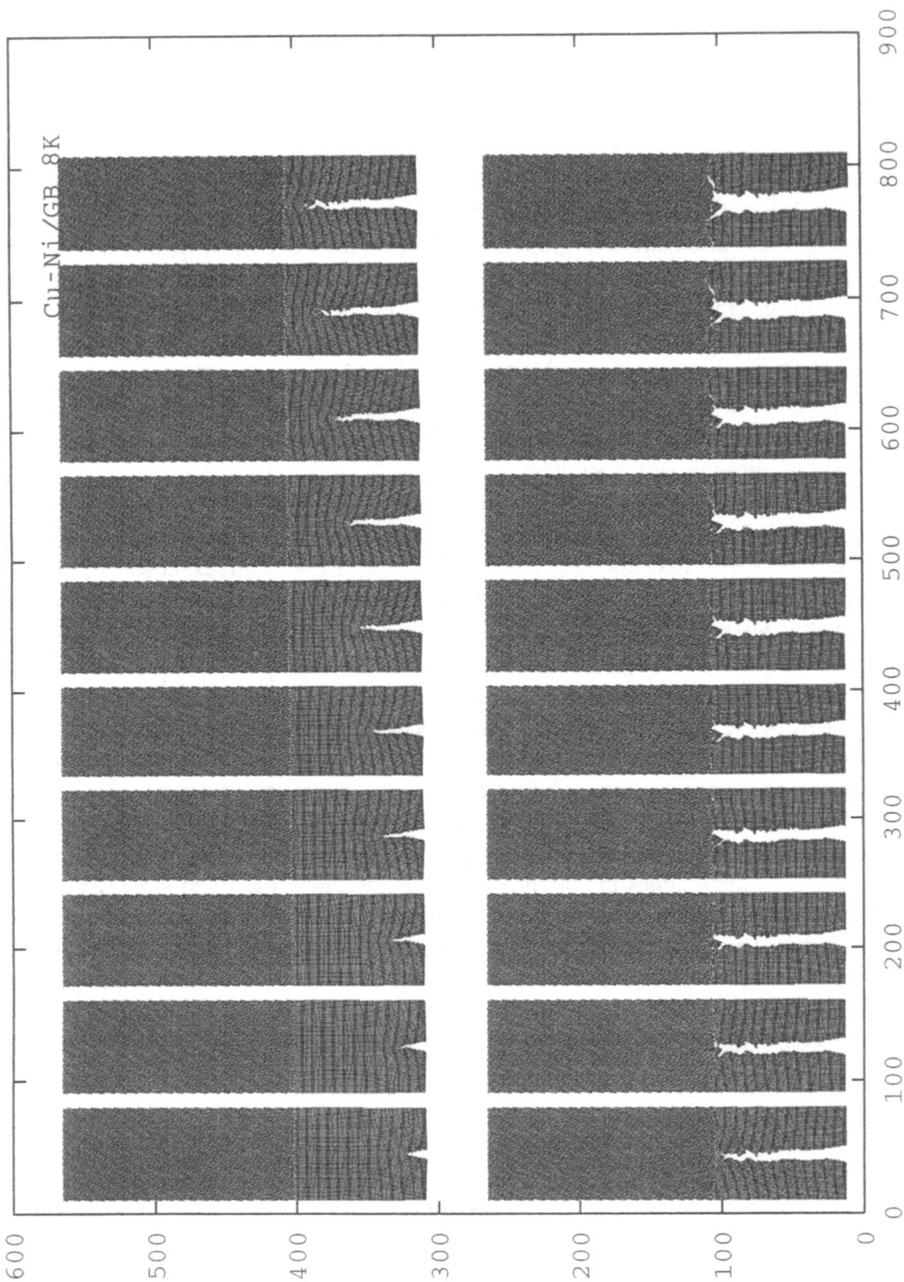

Fig. 4. Series of snapshots showing crack in Cu-Ni Monel alloy in the presence of a Σ-5 grain boundary. Notice the slowing down and blunting of the crack tip as it approaches the grain boundary and the creation of voids as the crack arrives at the boundary. First snapshot at 180 ps, time between snapshots 30 ps.

and growth. Many questions remain regarding the role of defects, the spatial extent of the system, the proper thermodynamic equilibrium, etc. Moreover, all simulations to date assume that a crack is already present in an otherwise fairly homogeneous material. The important technological questions of crack initiation and the related issue of metal fatigue [2] have not been addressed at all by the large-scale MD methods. At present, the time- scales on which these phenomena operate are simply too large for even the fastest parallel computers. Most of the results described here have been for metallic systems, although other materials have recently been studied by essentially the same methodology. Two recent examples include nanophase Silicon Nitride [35] and graphite [36].

One of the shortcomings of the current state-of-the-art in atomistic models of crack propagation is the use of classical potentials to describe processes (bond breaking, phonon generation, etc.) that are inherently quantum mechanical in nature. Clearly, the scale of fracture simulations, with millions of atoms, prohibits a full quantum mechanical treatment of the problem. However, an interesting possibility would be to treat a zone of atoms quantum mechanically and then join these results onto a classically modeled region. Possibilities for quantum mechanical simulation include tight-binding molecular dynamics (TB-MD) [48, 49] or the Car-Parrinello (C-P) method [50, 51]. The TB-MD method is a semi-empirical approach since it involves parameters that are fit to experiment or theoretically calculated band-structures. Nevertheless, its properties and shortcomings are well understood and, because it is fast, it allows fairly large systems to be treated. The C-P method on the contrary is an *ab initio* treatment, potentially the most accurate approach available, but computationally very demanding. A program based on combining tight-binding and classical simulations is currently being pursued by a consortium known as the MAAD (Macro-Atomistic *Ab-initio* Dynamics) coalition [52]. This group proposes to treat the majority of the system in a continuum approach, except for the vicinity of the crack which is modeled by a classical pair potential, with the tip of the crack where bond-breaking occurs treated quantum mechanically. This is a promising avenue of research.

The question of appropriate boundary conditions to absorb the outgoing shock wave remains a crucial one. It seems likely that further progress is needed to deal with this important consideration. The problem is, of course, ubiquitous in the modeling of impact phenomena. Recently, Haberland *et al.* [53], in simulations of small and medium-size clusters impacting a substrate, proposed a new way of dealing with the shock wave problem. They suggested to embed the discrete model in a continuum medium, with a transition zone in between, specifically designed to absorb shock waves, but mimicking the essential properties of the system being simulated. The idea of embedding a crack in some sort of a medium is not a recent one although it was originally mainly studied in the context of continuum elasticity theory [54]. There are further advantages to such an embedding approach: not only does it allow for a natural merging of the atomistic and continuum descriptions permitting the calculations to focus on the main region of interest, i.e. the vicinity of the crack tip in the case of fracture, it also drastically reduces the amount of data generated.

Fracture is a phenomenon that straddles a wide range of time and length scales. There is a clear need for a formulation that merges the discrete (microscopic) and continuum (macroscopic) description. Much research in industry uses a continuum approach, based on variations of such techniques as the Finite Element Method [55, 56]. If the academic research is to make an immediate impact on technology one of its aims should be to provide parameters, obtained from the atomistic simulations, that can be input into the continuum programs. Related to this is the problem of time scales. The atomistic simulations at best operate on the scale of ns, whereas many aspects of fracture occur on a much larger scale. Some way to speed up the simulations, perhaps by including only 'relevant' processes (bond breaking, dislocation emission, etc.), would be extremely useful. Of course, such a method does not currently exist. However, very recently Voter [57] has proposed a 'hyperdynamics' technique which is based on an accelerated MD in the case of infrequent events. It is not clear if this framework can be applied to the case of fracture, but it is an interesting possibility that seems well worth pursuing.

As this brief review has shown, realistic atomistic studies of fracture mechanics have become a reality. Although they remain at the cutting edge of what is computationally feasible, the anticipated improvements in hardware and software will undoubtedly push the field forward. Simulations on two-dimensional systems gave useful insights, but it is now generally recognized that a three-dimensional simulation is necessary to obtain realistic results, particularly relating to the emission of dislocations. Dimensions of the systems modeled are rapidly approaching the actual size of the microstructures in real solids [58]. However, progress will most likely not come from simulating increasingly larger systems, but rather from combined models that embed a realistically (perhaps quantum-mechanically) modeled micro-region in a macroscopic continuum. Exactly how to do this in a consistent fashion remains an unsolved problem. Likewise, a daunting problem remains in bridging the gap in time scales between simulations and crack initiation processes occurring in nature. Moreover, previous studies have mainly concentrated on perfect crystals, but the time has come to study the important effects of lattice imperfections, such as vacancies, interstitials, internal and external interfaces, impurities, etc. Finally, atomistic studies so far have only considered a single starter notch, but interesting effects may emerge when several microcracks are present simultaneously. Clearly, much has already been accomplished in atomistic fracture modeling, but many challenges remain and this problem promises to be in the forefront of computational materials science for a long time to come.

Acknowledgements

The author is greatly indebted to Dr W. C. Morrey for his assistance in some of the work that is discussed and reproduced here and for many enlightening discussions. Dr Morrey's work benefited greatly from a generous allocation of computer time by Dr C. Halloy at the Joint Institute for Computational Science at the University of Tennessee. Stimulating discussions with Drs B. L. Holian, R. Blumenfeld and F. F. Abraham are also gratefully acknowledged. A special thanks is due to Dr Abraham for permission to reproduce Fig. 3 and to Drs F. Cleri and S. Yip for making (p)reprints available. Finally, a sincere thanks goes to Professor Y. Kawazoe for the invitation to contribute this paper and for his patience in awaiting its completion.

References

1. L. B. Freund: *Dynamic Fracture Mechanics* (Cambridge University Press, Cambridge, 1990).
2. S. Suresh: *Fatigue of Materials* (Cambridge University Press, Cambridge, 1991).
3. R. H. Jones, W. W. Gerberich (eds): *Modeling Environmental Effects on Crack Growth Processes* (The Metallurgical Society, Warrendale, PA, USA, 1986).
4. O. Buck (ed.): MRS Bull. **14**, No. 8 (1989) - special issue on *Crack formation and propagation.*
5. H. J. Herrmann, S. Roux (eds.), *Statistical Models for the Fracture of Disordered Media* (North-Holland, Amsterdam, 1990).
6. M. Marder, J. Fineberg: Physics Today **49**, No. 9, 5 (1996).
7. P. Meakin: Science **252**, 226 (1991).
8. J. R. Rice, R. M. Thomson: Philos. Mag. **29**, 73 (1974).
9. S. J. Zhou, A. E. Carlsson, R. Thomson: Phys. Rev. Lett. **72**, 852 (1994).
10. A. A. Griffith: Philos. Trans. R. Soc. Lond. **A 221**, 163 (1920).

11. S. Ramanathan, D. S. Fisher: Phys. Rev. Lett. **79**, 877 (1997); and references therein.
12. B. L. Holian, R. Blumenfeld, P. Gumbsch: Phys. Rev. Lett. **78**, 78 (1997).
13. S. J. Zhou, N. Gronbech-Jensen, A. R. Bishop, P. S. Lomdahl, B. L. Holian: Phys. Lett. A **232**, 183 (1997).
14. W. G. Hoover: *Molecular Dynamics* (Springer-Verlag, Berlin, 1986).
15. G. Ciccotti, D. Frenkel, I. R. McDonald: *Simulation of Liquids and Solids* (North-Holland, Amsterdam, 1987).
16. M. P. Allen, D. J. Tildesley: *Computer Simulation of Liquids* (Oxford University Press, Oxford, 1989).
17. R. P. Feynman, R. B. Leighton, M. Sands: *The Feynman Lectures on Physics*, Vol. 1, Section 9 (Addison-Wesley, Reading, 1963).
18. B. J. Alder, T. E. Wainwright: J. Chem. Phys. **27**, 1208 (1957); reprinted in Ref. 15.
19. An excellent overview of the state-of-the-art in the construction of interatomic potentials may be found in the February 1996 issue of MRS Bull. (Vol. **21**, No. 2), edited by A. F. Voter.
20. M. S. Daw, M. I. Baskes: Phys. Rev. B **29**, 6443 (1984); M. S. Daw, S. M. Foiles, M. I. Baskes: Mater. Sci. Rep. **9**, 251 (1993).
21. K. S. Cheung, S. Yip: Phys. Rev. Lett. **65**, 1804 (1990); Model. Simul. Mater. Sci. Eng. **2**, 865 (1994).
22. B. L. Holian, A. F. Voter, N. J. Wagner, R. J. Ravelo, S. P. Chen, W. G. Hoover, C. G. Hoover, J. E. Hammerberg, T. D. Dontje: Phys. Rev. A **43**, 2655 (1991).
23. N. J. Wagner, B. L. Holian: Mater. Res. Soc. Symp. Proc. **291**, 91 (1993).
24. B. L. Holian, R. Ravelo: Phys. Rev. B **51**, 11275 (1995).
25. S. J. Zhou, D. M. Beazley, P. S. Lomdahl, B. L. Holian: Phys. Rev. Lett. **78**, 479 (1997).
26. N. J. Wagner, B. L. Holian, A. F. Voter: Phys. Rev. A **45**, 8457 (1992).
27. F. F. Abraham, D. Brodbeck, R. A. Rafey, W. E. Rudge: Phys. Rev. Lett. **73**, 272 (1994).
28. F. F. Abraham: Phys. Rev. Lett. **77**, 869 (1996); Europhys. Lett. **38**, 103 (1997); IEEE Comput. Sci. Eng. **4**, No. 2, 66 (1997).
29. W. C. Morrey, L. T. Wille: Mater. Sci. Eng. B **37**, 181 (1996); Mater. Res. Soc. Symp. Proc. **409**, 81 (1996).
30. W. C. Morrey: *Massively Parallel Molecular Dynamics Simulation of Crack Initiation and Growth in Copper-Nickel Alloys* (Ph. D. dissertation, Florida Atlantic University, 1996).
31. S. J. Zhou, P. S. Lomdahl, R. Thomson, B. L. Holian: Phys. Rev. Lett. **76**, 2318 (1996).
32. P. Gumbsch, S. J. Zhou, B. L. Holian: Phys. Rev. B **55**, 3445 (1997).
33. W. T. Ashurst, W. G. Hoover: Phys. Rev. B **14**, 1465 (1976).
34. R. A. Johnson: Phys. Rev. A **134**, 1329 (1964).
35. R. K. Kalia, A. Nakano, A. Omeltchenko, K. Tsuruta, P. Vashishta: Phys. Rev. Lett. **78**, 2144 (1997).
36. A. Omeltchenko, J. Yu, R. K. Kalia, P. Vashishta: Phys. Rev. Lett. **78**, 2148 (1997).
37. D. M. Beazly, P. S. Lomdahl: Parallel Comput. **20**, 173 (1994).
38. K. M. Nelson, C. F. Cornwell, L. T. Wille: Comput. Mater. Sci. **2**, 525 (1994).
39. S. Plimpton: J. Comput. Phys. **117**, 1 (1995).

40. Y. S. Hwang, R. Das, J. H. Saltz, M. Hodoscek, B. R. Brooks, Comput. Sci. Eng. **2**, No. 2, 18 (1995).
41. L. T. Wille, C. F. Cornwell, W. C. Morrey: Mater. Res. Soc. Symp. Proc. **408**, 125 (1996).
42. G. C. Fox, M. A. Johnson, G. A. Lyzenga, S. W. Otto, J. K. Salmon, D. W. Walker: *Solving Problems on Concurrent Processors* Vol. I (Prentice Hall, Englewood Cliffs, 1988), pp. 203-211.
43. B. L. Holian, O. E. Percus, T. T. Warnock, P. A. Whitlock: Phys. Rev. E **50**, 1607 (1994).
44. D. M. Beazley, P. S. Lomdahl: Comput. Phys. **11**, 230 (1997).
45. S. Yip, D. Wolf: Mater. Sci. Forum **46**, 77 (1989).
46. M. P. Seah: J. Phys. F, Met. Phys. **10**, 1043 (1980).
47. S. M. Foiles: Phys. Rev. B **40**, 11502 (1989).
48. K. Laasonen, R. M. Nieminen: J. Phys., Condens. Matter **2**, 1509 (1990).
49. S. Goedecker, M. Teter: Phys. Rev. B **51**, 9455 (1995).
50. R. Car, M. Parrinello: Phys. Rev. Lett. **55**, 2471 (1985).
51. M. C. Payne, M. P. Teter, D. C. Allan, T. A. Arias, J. D. Joannopoulos: Rev. Mod. Phys. **64**, 1045 (1992).
52. F. F. Abraham: Phys. Today **50**, No. 2, 15 (1997).
53. H. Haberland, Z. Insepov, M. Moseler: Phys. Rev. B **51**, 11061 (1995).
54. F. Cleri, S. R. Phillpot, D. Wolf, S. Yip: in *Advances in Computational Materials Science* (eds V. Fiorentini and F. Meloni), Vol. 55, (SIF, Bologna, 1997).
55. P. Gumbsch: J. Mater. Res. **10**, 2897 (1995).
56. P. Gumbsch, G. E. Beltz: Model. Simul. Mater. Sci. Eng. **3**, 597 (1995).
57. A. F. Voter: Phys. Rev. Lett. **78**, 3908 (1997).
58. S. S. Hecker: MRS Bull. **XX**, No. 4, 4 (1995).

Part 2

Deformation and Fracture

Nanoscopic Modelling of the Adhesion, Indentation and Fracture Characteristics of Metallic Systems via Molecular Dynamics Simulations

H. Rafii-Tabar

Nano-Science Simulation Group, Centre for Numerical Modelling & Process Analysis, School of Computing & Mathematical Sciences, University of Greenwich, Woolwich Campus, Wellington Street, London SE18 6PF, UK

Summary. Large-scale molecular dynamics simulations have been performed on canonical ensembles to model the adhesion and indentation characteristics of 3-D metallic nano-scale junctions in tip-substrate geometries, and the crack propagation in 2-D metallic lattices. It is shown that irreversible flows in nano-volumes of materials control the behaviour of the 3-D nano-contacts, and that local diffusional flow constitutes the atomistic mechanism underlying these plastic flows. These simulations show that the force of adhesion in metallic nano-contacts is reduced when adsorbate monolayers are present at the metal–metal junctions. Our results are in agreement with the conclusions of very accurate point-contact experiments carried out in this field. Our fracture simulations reveal that at low temperatures cleavage fractures can occur in both an elemental metal and an alloy. At elevated temperatures, the nucleation of dislocations is shown to cause a brittle-to-ductile transition. Limiting crack propagation velocities are computed for different strain rates and a dynamic instability is shown to control the crack movement beyond this limiting velocity, in line with the recent experimental results.

Keywords. Nano-contact modelling, MD simulation, crack propagation, tip-substrate interaction, adhesion, indentation

1. Introduction

Nano-science and nano-technology [1] now form two major pathways in the basic research concerned with the purposeful manipulation and structural modification of materials on atomic and molecular scales, and with the further progress of the *miniaturization programme* aimed at the design and fabrication of components, such as low-dimensional quantum devices, operating on significantly reduced length and time scales as compared with the existing devices.

In materials science and condensed matter physics research, and more generally in the field of nano-science research, detailed knowledge of the deformation and response characteristics of *nano-structures* can provide deep insights into the microscopic basis of many materials phenomena on macroscopic scales, and in recent years this has prompted a considerable interest in the atomistic modelling of interfacing nano-crystals of solids, particularly of metallic and semiconducting materials, as well as of the crack propagation in elemental and alloyed nanostructures. The modelling of these *nano-contacts* is seen as an elegant method to reveal the fundamental physical mechanisms that contribute to the onset of *plastic flow* in ultra-fine volumes of solids. This nano-flow is considered as the process ultimately responsible for the *adhesion, cohesion, friction, indentation, fracture* and *surface modification* of material systems on *macroscopic scales*, as well as for their substantial modification when molecularly-thick adsorbate layers, such as solid or fluid films, are introduced at the interfaces. This emerging field of research has been much aided by the complementary experimental studies [2, 3] designed to measure directly the local (nano-scale) mechanical properties, such as the indentation hardness and elastic constants.

Atomistic modelling of the adhesive and indentation properties of materials presupposes a radical modification of our concept of a *material surface* as a simple geometric boundary separating two condensed phases. Close inspections of solid surfaces by local scanning probe devices reveal that, even after the most careful finishing, these surfaces are "rough" on atomic and molecular scales. The roughness is generated by *random fluctuations* in the surface, of short wavelength, and is characterised by "atomic hills" (nano-scale asperities) and "valleys" of varying amplitudes and spacings [4]. These irregularities are *stochastic structures* on a nanoscopic scale consisting of hundreds of atoms and molecules. The distribution of such atomic-scale asperities over a surface can be either directional or homogenous in all directions, depending upon the nature of the processing method involved. *We can therefore conclude that at the nanoscopic level of description, the surfaces of materials, particularly of metals, can have very complex morphologies consisting of multitudes of nano-particles, and that their physical attributes and properties are far more complex than those implied by the simple continuum-scale model.*

When two extended solid surfaces with such nano-scale features are brought into close proximity, their apparent area of contact is significantly different from that of a real contact. Intimate contact takes place where the tips of nano-asperities on one surface touch those on the other. The energetics of these individual asperities, therefore, determine the sizes of the regions where adhesion, surface friction and possible interfacial cohesion take place, and how one loading solid can induce indentation and fracture in another solid below. Furthermore, the degree of roughness of a surface also influences the kinetics of its wetting by adsorbates and may change the apparent advancing contact angle for a given liquid on a rough solid [4].

Once an interfacial contact is made the asperity regions initially deform elastically and, under sufficient loads, plastic deformation or strain is then produced in the asperities themselves until the load can be supported.

The deformations occur in the cohesively weaker of the two materials in contact. For most practical materials, the asperity contact (even under the lightest of loads) experiences a very high local stress in the real area of contact of the asperities, which is generally sufficient to produce strain in the asperities near the tips [4]. This means that the real contact area continues to grow.

There exist several continuum-elasticity and contact-mechanics models of interfacing surfaces which can be used to estimate the area of contact in terms of the elastic and plastic properties of surfaces [5, 6]. *However, since surface asperities exist on nanoscopic scales, their elastic, plastic and fracture properties differ from those in bulk, as the mechanical properties show a strong dependence on the sample sizes.* This implies that the continuum-based elasticity theories are not appropriate to nano-contacts, and that we need to model these at the nano-scale.

A significant development in our understanding of the deformation and adhesive processes at nano-contacts, particularly of metallic junctions, has been brought about by *point-contact* experimental studies concerned with exploring the local properties of small volumes of materials using tip-based probe devices akin to the scanning tunnelling microscope (STM) and the atomic force microscope (AFM). In two such experiments [2, 3] a *sharp tip*, mimicking the geometry of a nano-asperity, was brought into contact with a *flat substrate* and was then either allowed to pass the point of contact [3], resulting in a nano-indentation, or pulled off the substrate to measure the force of a clean adhesive contact as well as its modification due to oxygen monolayers adsorbed onto the substrate [2].

From these two typical experiments we can draw the following general conclusions: both indentation and adhesive contact of the surfaces were clearly dominated by the onset of plastic flows in ultra-fine volumes of materials as evidenced by the observation of pronounced hystereses in the load/unload cycles. This type of behaviour was observed even at applied loads so small that the deformation was dominated by the action of the surface forces. *Furthermore, in the adhesion experiment [2], the force of adhesion, i.e the negative force required to separate the surfaces, was significantly reduced when monolayers of oxygen were introduced at the interface.*

The tip-substrate 3-D modelling studies reported here were concerned with testing these general conclusions for interfacing metallic nano-crystals. *In particular we aimed to obtain the mechanisms contributing to the onset of plastic flow processes during these nano-contacts.*

Numerical modelling of *crack propagation* in metals is another aspect of the research reported in this paper. Traditionally, such modelling has been performed on the basis of continuum-elastic, or elastic-plastic, theories of fracture mechanics using a variety of energy and force criteria [7]. These theories describe the conditions for propagation of an already existing crack on the verge of extension. Despite their undoubted successes, modelling based on these theories is not capable of predicting all of the observed characteristics of crack propagation or crack geometries. For example, whereas continuum-based theories assume *smooth* crack surfaces and obtain a crack *limiting velocity* equal to the Rayleigh wave speed of the material [7], V_R, recent experimental studies [8, 9] clearly indicate that crack surfaces that are smooth, or *mirror-like*, to begin with, undergo *mirror-to-mist-to-hackle* roughening transitions, and that crack velocities reach limiting values of about 0.6 V_R. Furthermore, these experiments indicate that a *dynamic instability* controls the crack's advance beyond a velocity of 0.36 V_R, leading to *violent velocity oscillations* that correlate strongly with the emergence of roughening transitions of the crack surfaces.

These phenomena ultimately depend on a variety of fundamental processes that occur within highly localised *ultra-fine* volumes of materials that form the nano-scale *fracture process zones* surrounding the crack tips, and therefore a nano-scale modelling of crack propagation in this zone will be able to provide deep insights into the origins of these crack properties. We adopted such an approach to model the crack propagation in a 2-D Ag lattice and a 2-D lattice of Ag-Au random binary alloy.

In Sections 2, 3 and 4, following a brief discussion of the molecular dynamics (MD) simulation methodology of canonical ensembles, we will present the results of our modelling of the indentation and adhesive properties of solid-solid and solid-monolayer-solid junctions involving tip-substrate geometries, and of the crack propagation in 2-D metallic lattices.

2. Molecular Dynamics Simulations

Computer-based Molecular Dynamics (MD) simulation [10] has now become the most significant methodology to study the spatio-temporal evolution of many-body systems in condensed matter physics and materials science at length and time scales that may not be accessible to current experimental techniques.

The MD simulations, performed at the atomistic and molecular levels, are capable of providing deep insights into the underlying processes that control the structure and collective behaviour and properties of complex systems and their emergent properties.

A classical MD simulation proceeds by considering a nano-sized model-system of a condensed phase in a primary computational cell which is periodically replicated in three dimensions to minimise the effects of the artificial surfaces that are introduced due to the finite size of the model. The forces experienced by the particles of the system are derived from two-body, and many-body, inter-atomic potentials, and are used in the coupled equations of motion, in both deterministic (Newtonian) and stochastic (Langevin) dynamics, to provide the space-time trajectories of the particles. Powerful statistical-mechanics methods, such as the fluctuation theories of statistical ensembles, are used to relate the time-averaged microstate information generated along these trajectories via MD to the ensemble-averaged static and dynamic properties of the system.

2.1 Inter–atomic Potentials

The use of inter-atomic potentials lies at the very foundation of MD simulations, and the significance and reliability of much of the simulation results depend crucially on the correct modelling of the energetics of the particles involved.

In the MD simulations reported here, the inter-atomic forces for both our 3-D tip-substrate and 2-D crack propagation modelling were derived from the following *long-range, many-body* potential [11] that had been specifically developed for the atomistic modelling of the fcc random binary alloys

$$H = \sum_i \sum_{j>i} V(r_{ij}) - d^{AA} \sum_i \hat{p}_i \sqrt{\rho_i^A} - d^{BB} \sum_i (1 - \hat{p}_i) \sqrt{\rho_i^B} \qquad (2.1)$$

providing accurate results on lattice parameters, elastic constants and enthalpies of formation of these alloys. The specific forms of the various terms are given in [11].

The Hamiltonian in (2.1) is an inter-atomic potential of the Finnis–Sinclair type [12] and is a generalisation of the potential [13] describing the fcc elemental metals. It allows us to model both the short- and the long-range contributions to the atomic interactions, over a wide range of distances, in an MD simulation involving clusters of unlike atoms. It combines the van der Waals attractions at the long range with the cohesive interactions at the short range and has the added advantage that all of the parameters of the A-B interactions can be obtained from those relevant to the A-A and B-B interactions *without* the need for the introduction of any further parameters. This Hamiltonian has formed the basis of a number of large scale MD simulations [14–19].

From this Hamiltonian, the forces acting on each atom were computed and then used in the equations of motion based on the velocity Verlet algorithm [10] to produce the trajectories of individual atoms. For the 3-D modelling, the equations of motion were integrated in a 3-D simulation cell with periodic boundary conditions (PBC) applied in all directions. The image cells contained the periodic images of the atoms in the central cell, and the periodic minimum image interactions [10] were also used. For the 2-D simulations, no PBCs were applied.

2.2 Control of Simulation Temperature

The classical MD simulations reported here were of the equilibrium type performed on systems in which the total number of particles (N), the volume of the system (V) and the temperature of the system (T) were all kept at constant preset values. In statistical mechanics, the properties of such systems are studied within the framework of a constant-(NVT) *canonical ensemble* [20] in which the temperature, rather than the total energy, acts as a control parameter. The temperature is maintained at a constant value by surrounding the physical system with an external *heat bath* that exchanges energy, but not particles, with the system. The temperature T corresponds to the temperature of this heat bath.

There exist several methods of implementing a constant temperature simulation. In all these methods [10], a thermal contact is implicity assumed between the system and a heat bath. The distinguishing feature of each method is the *physical mechanism* by which this contact is established. In our tip-substrate simulations, we implemented a canonical ensemble by adopting the Nosè–Hoover (NH) *thermostat* formalism [21]. This formalism is based on an *extended system* method in which the physical system is connected to a thermostat whose function is mathematically represented by an additional degree of freedom coupled to the kinematic variables of the system and which acts as a scaling parameter. The atoms in the system are then either accelerated or decelerated depending on whether the *time-dependent* friction coefficient of the thermostat becomes, respectively, negative or positive.

For the crack propagation simulations, the temperature was fixed by rescaling the velocities of the atoms at each time step.

2.3 Tip–Substrate Simulation Method and Data

In our MD simulations we considered an atomically-sharp tip coming into contact with a flat substrate. We then either allowed the tip to pass the point of contact and indent the substrate, or pulled the tip off the surface to measure the force of adhesion when the surface was clean as well as when it was covered with either a soft or a hard monolayer of an intervening material.

The motion of a tip relative to a substrate can be modelled in two different ways. A simple standard way is to make certain atomic layers in both the tip and the substrate, far from the region of their mutual interactions, static and then move these static layers relative to each other. This method leads to the formation of rather artificial interfaces which may act to reflect thermal vibrations as well as mobile defects. The other method, adopted in our simulations, is to use the PBC and consider two cells containing the tip-substrate system and its periodic image. Then by changing the height of one cell in the z direction, the tip in that cell can be made to move relative to the substrate in the other (periodic image) cell directly below. In this way, the substrate acts both as the tip holder and the surface with which the tip interacts during the simulation runs.

The substrate in each simulation had the dimensions $11a \times 11a \times 3a$ (where a is the fcc lattice parameter) consisting of 7 layers of *dynamic* atoms with 242 atoms per layer. The sharp tip was constructed by placing the atoms of an fcc crystal structure into a paraboloid of radius of curvature of $1a$ and the height of $5a$. The tip consisted of 273 atoms, giving a total of 1967 atoms in each simulation. The substrate normal was (001). The potential was truncated at a distance of $2a$ giving the interactions of 140 neighbours on each atom considered. The simulation temperature was 300 K and the time step for all simulations was 10^{-14} s.

A set of five tip-substrate simulations involving a soft material (Pb) and a hard material (Ir) were performed [14–16]. We selected these particular metals for their very diverse mechanical and thermal properties. Ir is some nine times harder than Pb. For all the simulations, an equilibration run at constant temperature and pressure was first performed during which the tip was kept sufficiently far from the substrate so that there were no interactions between the two. For the indentation case, the equilibration run consisted of 4000 time steps, and for the adhesion cases these runs were of 2000 time step duration.

3. Tip–Substrate Simulation Results

3.1 Nano–indentation of a Soft Substrate by a Hard Tip

Following equilibration, the Ir tip was lowered toward the Pb substrate in the manner discussed above. Figure 1 shows the 3-D snapshots of the system during a complete cycle of approach, contact, indentation and retraction of the tip from the substrate. The retraction of the tip from the substrate led to the formation and elongation of an atomic neck. The variations of the computed force on the lowermost tip atom versus the tip-to-substrate distance are shown in Fig. 2 for the complete cycle. On the approach part of this figure we observe the phenomenon of *jump-to-contact* (JC) [22], signalled by an increase in the attractive force between the tip and the substrate (point A1). During the JC instability, the Pb atoms directly below the tip were displaced toward the tip resulting in a local deformation (bulging) of the substrate surface(Fig. 1c).

The sharp fluctuations observed in the segment BC correlate strongly with the extensive structural deformations in the substrate, as shown in Fig. 1d. The hysteresis observed in Fig. 2 is a clear indication of the onset of plastic deformation in the substrate. From the 3-D geometries we can observe that this deformation was brought about by the local diffusional flow of the substrate atoms. *We can therefore suggest that local diffusional flow is an atomistic mechanism contributing to the onset of irreversible deformations in defect-free ultra-fine volumes of materials.* The local diffusion led to a 'pile up' of the substrate atoms around the edges of the tip producing the *wetting* of the tip by these atoms. During the entire simulation,

the tip preserved its structural integrity without any of its atoms diffusing into the substrate.

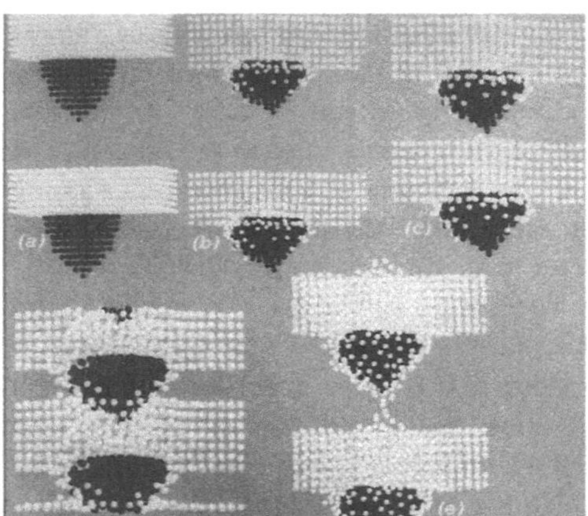

Fig. 1. Three-dimensional geometries of the tip-substrate system (**a**) initial state; (**b**) after equilibration; (**c**) after jump-to-contact; (**d**) at maximum indentation; (**e**) retraction showing the formation of an atomic neck.

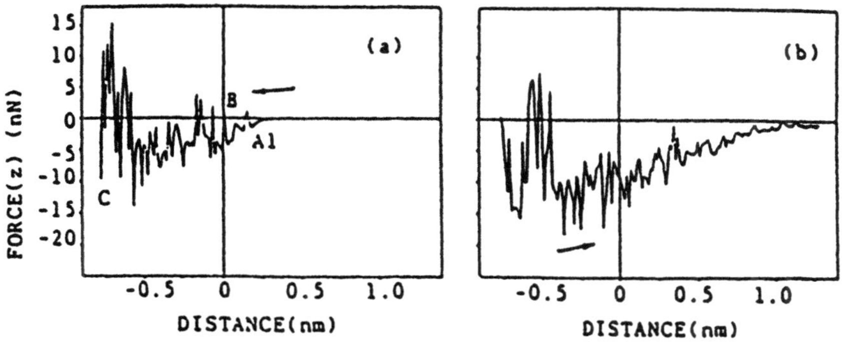

Fig. 2. Calculated force on the tip atom in z direction versus tip-to-substrate distance for (**a**) approach, jump-to-contact and indentation; (**b**) retraction.

3.2 Adhesion of Surfaces

In these simulations, following the equilibration phases, the tip was brought into adhesive contact with the substrate and then its direction was reversed to measure the negative force necessary to separate the surfaces, i.e. the force of adhesion.

Figure 3 shows the variations of this force with the tip-to-substrate distance for clean surfaces, i.e. the Pb/Pb and Ir/Ir junctions, and in Fig. 4 we give, as an example, a snapshot of the Pb tip being pulled off the Pb substrate where the interfacial region of the tip is seen to have undergone large morphological transformations resembling a liquid drop. During the pull-off, we again observed the necking down of the tip as it was separated from the substrate, and some tip atoms were transferred to the surface.

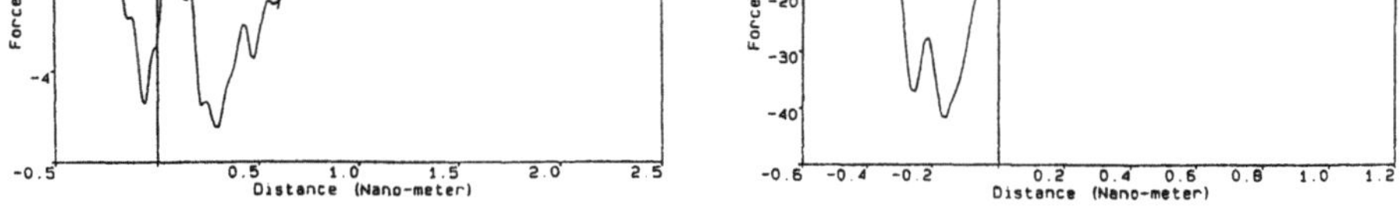

Fig. 3. Variations of the force of adhesion versus distance for (**a**) Pb/Pb junction; (**b**) Ir/Ir junction.

In a similar simulation involving the Ir/Ir junction, we again observed that the interfacial part of the tip became highly disordered forming a relatively small neck during the pull-off, and again some atoms were left behind. However, since a smaller part of the Ir tip was involved in the formation of the neck, fewer atoms were left on the substrate. This is reflected in the smaller contribution of the tail segment in the force-displacement figure when it is compared with the Pb/Pb case.

The pronounced hystereses observed in Fig. 3 again suggest the onset of irreversible flows in these clean-surface adhesive contacts. Moreover, the dominant mechanism here is again a diffusional creep which is driven by the stresses acting in the tip. No indentations in these simulations were observed since the tip in each case deformed and relieved the applied stresses.

Figure 5 shows a snapshot of the Pb/Ir/Pb junction. The Ir monolayer was introduced by converting the top layer of the Pb substrate. However, after an equilibration run of 1000 time steps it diffused into the substrate. We therefore increased the thickness of the substrate by an extra layer. The JC phenomenon in this case was marked by the atoms at the base of the Pb tip straining toward the Ir film. This is because the Ir–Pb bond is far stronger than the Pb–Pb bond. This led to a very rapid formation of a large contact area. The tip appeared to have melted and taken on the structure of a liquid drop wetting the Ir film. During the tip pull-off, we observed the necking down of the tip, leaving more Pb atoms on the Ir than in the previous case (Fig. 4). This was due to a larger contact area.

A similar simulation was performed for the Ir/Pb/Ir junction. In this case, the JC was marked by the Pb atoms in the film straining upward and wetting the tip. Since the Ir–Ir bonds are stronger than the Pb–Ir bonds, no Ir atoms were left on the Pb film upon the retraction of the tip. The main effect of the Pb film was to prevent the direct formation of the Ir–Ir bonds between the tip and the substrate.

Figure 6 shows the variations of the force of adhesion for these two cases. We can clearly see that this force is reduced when compared with the corresponding clean-surface contact. *From these adhesive simulations we can conclude that irrespective of whether wetting between the tip and substrate is very favourable, as*

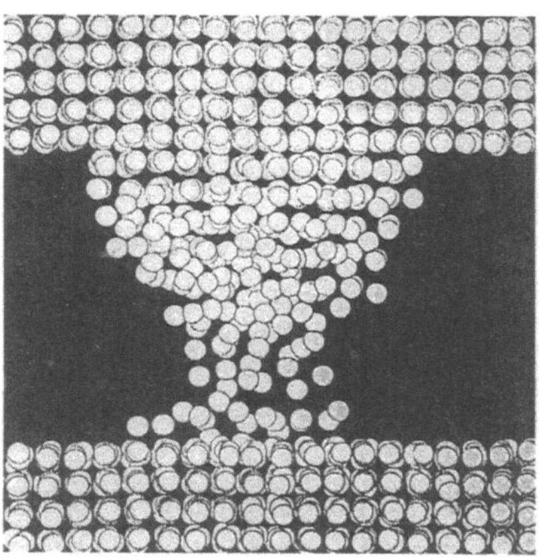

Fig. 4. Pb tip being pulled off the Pb substrate.

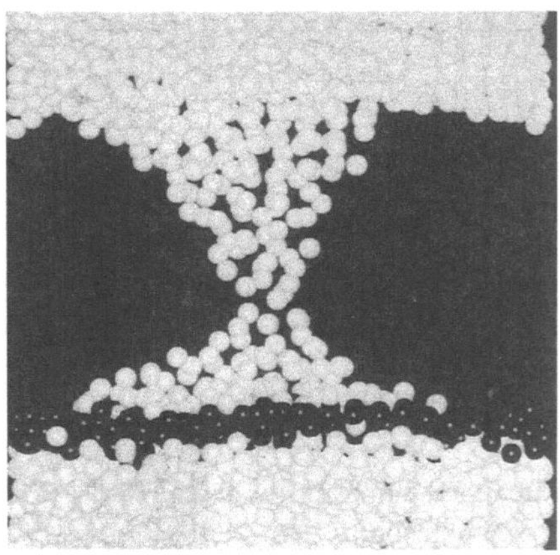

Fig. 5. Pb tip being pulled off the Pb substrate with an intervening Ir monolayer present.

in the Pb/Ir/Pb case, or not favourable, as in the Ir/Pb/Ir case, the force of adhesion is always reduced relative to the clean surfaces. This is in agreement with the general conclusions reached experimentally [2], even though the materials employed in our simulations were different from those studied in that experiment.

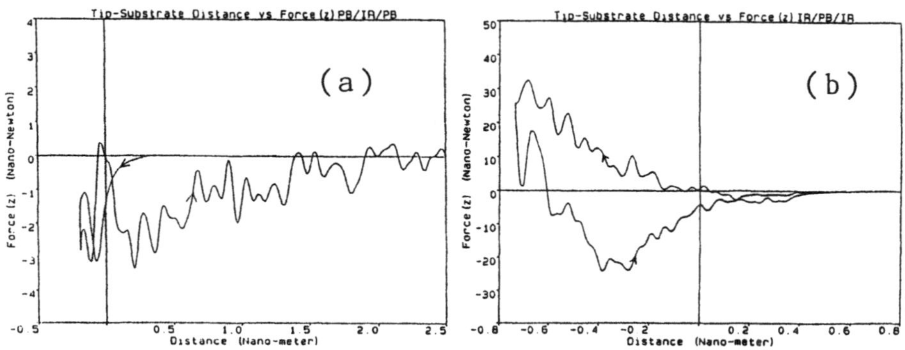

Fig. 6. Variations of the force of adhesion versus distance for (**a**) Pb/Ir/Pb junction; (**b**) Ir/Pb/Ir junction. Notice the reduction in force as compared with Fig. 4.

4. Simulation of Crack Propagation in 2-D Metallic Lattices

4.1 Simulation Method and Data

The energetics and dynamics of the Ag and Ag-Au systems were derived from the potentials given in [13] and (2.1) respectively with the cut-offs set at the third nearest neighbours. To follow the propagation of a crack, the crack-tip atom was identified as the *most stressed* atom in the lattice by computing the stress-tensor field at individual atomic sites [23] which, for the potential in (2.1), is given in [18]. The lattices in both systems were triangular, consisting of 4970 atoms arranged in 71 rows with 70 atoms per row. The Ag-Au lattice was composed of 20 at % Au and 80 at % Ag. The initial positions of the atoms were those of the equilibrium lattice at zero temperature. To simulate the fracture process, mode-I loading was applied with an initial crack of length 10 d(d = triangular lattice parameter) inserted. A tensile stress was imposed by the application of forces to the top and bottom of the lattice. The stress was expressed in terms of force per atom and the unit chosen was the pico Newton (pN / atom). The lattice boundaries were not constrained. The initial crack was inserted by removing the interactions between the atoms on both sides of the crack plane.

The cracked lattice was first brought to equilibrium under a stress of 136.4 pN / atom. This was just below the critical stress which was obtained as $\sigma_c = 136.8$ pN / atom. The stress rate was then increased with each MD time step, which was set at 0.01 ps. As the stress was increased, the crack began to propagate in the lattice. In the simulation of brittle fracture, the temperature was initially set at 1 K but was allowed to evolve freely. For the ductile fractures, the temperature was maintained at 200 K.

4.2 Crack Propagation Simulation Results

Figure 7a shows a snapshot of the Ag lattice with an initial crack with mirror-like surfaces. The initial temperature was set at 1 K, but allowed to evolve freely. It is evident that the crack was in a cleavage mode. However, as the crack elongated and the crack velocity increased, new crack tips were nucleated and the original crack underwent a *bifurcation, or branching transition.* Figure 7b shows the crack after branching. It is clear that the original mirror-like surfaces have undergone roughening transitions. In Fig. 8a we show the variations of the crack length up to the first bifurcation point, and Fig. 8b shows the variations of the crack velocity. It is quite clear that just prior to branching the velocity was uniformly increasing to a maximum value of 2.1 d / ps, which was about 0.32 V_R, for all the different applied stress rates, and that following this limit strong oscillations have appeared. These results are in agreement both with the experimental findings [9] and the MD simulation performed on a *million-atom* system [24]. These oscillations strongly correlate with the mirror-to-mist-to-hackle transition shown in Fig. 7b.

Brittle fracture was also observed for the Ag-Au alloy at 1 K. The crack path in this case was more complicated than the Ag case due to the distortion of the lattice resulting from the presence of two types of atoms.

Figure 9 shows the development of ductile fractures in both the Ag and the alloy lattices at the constant elevated temperature of 200 K and the applied stress rate of 0.2 pN / atom/ ps. In both cases, dislocations were nucleated near the crack tip, blunting the tip and increasing the crack opening displacement (COD), and then migrating to the lattice boundary at an inclination of 60 degrees to the crack surface.

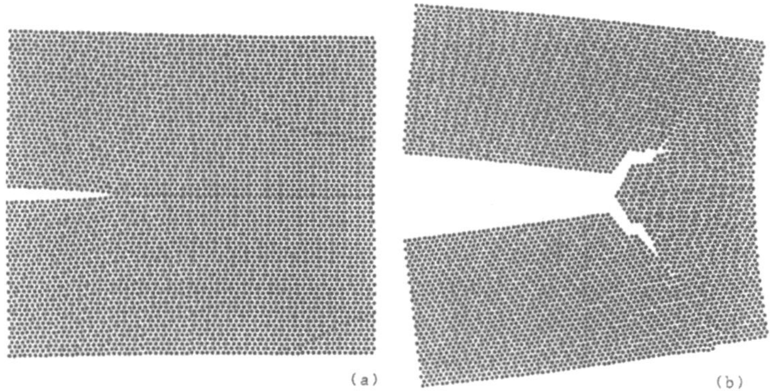

Fig. 7. Snapshots of the cracked Ag lattice showing (**a**) mirror-like surfaces; (**b**) after crack bifurcation leading to roughening transitions.

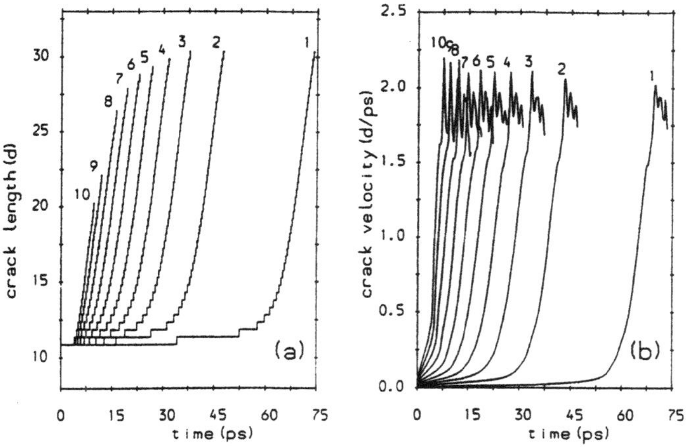

Fig. 8. Variations of (**a**) crack length and (**b**) crack velocity in the Ag lattice. Curves 1 to 10 correspond to stress rates from 0.025 to 12.8 pN / atom, doubling from one curve to the next.

The dislocation in the Ag case eventually left the lattice, whereas the dislocation in the alloy did not move out and also produced a vacancy as can be seen in Fig. 9b. *We can conclude that in our random binary alloy, the alloying had a small effect in hindering the dislocation mobility.*

In these extensive simulations we have been able to elucidate several important questions that can be of relevance to most research programmes within the emerging field of *nano-science simulation*. We have found that our simulations, based on our prescribed inter-atomic potentials, were capable of modelling the nano-scale adhesion, indentation and fracture characteristics of metallic systems correctly and shed light on some of the crucial questions concerning plastic flow processes and crack propagation in nano-volumes of materials.

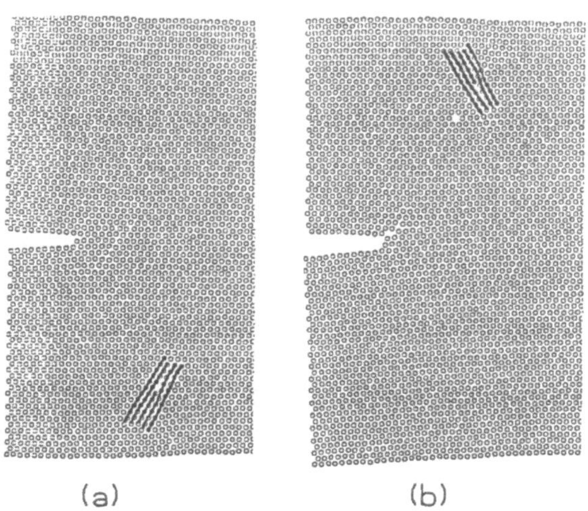

Fig. 9. Snapshots of (**a**) Ag and (**b**) Ag-Au lattices at 200 K showing dislocation motion.

References

1. K.E. Drexler, *Nanosystems* (J. Wiley Inc. N.Y, 1992).
2. M.D. Pashley, J.B.Pethica and D. Tabor: Wear **100**, 7 (1984).
3. J.B. Pethica and W.C. Oliver: Mat. Res. Soc. Symp. Proc. **130**, 13 (1989).
4. D.H. Buckley: *Surface Effects in Adhesion, Friction, Wear and Lubrication* (Elsevier, 1981).
5. T.R. Thomas: *Rough Surfaces* (Longmans, London, 1982).
6. J.A. Greenwood: Proc. R. Soc. London Ser. A **393**, 133 (1984).
7. L.B. Freund: *Dynamic Fracture Mechanics* (Cambridge University Press, 1990).
8. J. Finberg et al.: Phys. Rev. Lett. **67**, 457 (1991); Phys. Rev. B **45**, 5146 (1992).
9. E. Sharon et al.: Phys. Rev. Lett. **74**, 5096 (1995).
10. M.P. Allen and D.J. Tildesley: *Computer Simulation of Liquids* (Clarendon Press, Oxford 1987).
11. H. Rafii-Tabar, A.P. Sutton: Philos. Mag. Lett. **63**, 217 (1991).
12. M.W. Finnis, J.E. Sinclair: Philos. Mag. A **50**, 45 (1984).
13. A.P. Sutton, J. Chen: Philos. Mag. Lett. **61**, 139 (1990).
14. H. Rafii-Tabar et al.: Mat. Res. Soc. Symp. Proc. **239,** 133 (1991).
15. H. Rafii-Tabar, Y. Kawazoe: Jpn. J. Appl. Phys. **32**, 1394 (1993).
16. A.P. Sutton, J.B. Pethica, H. Rafii-Tabar, J.A. Nieminen: in "*Electron Theory in Alloy Design*" (Institute of Materials, London, 1994).
17. H. Rafii-Tabar et al.: Surf. Sci. **385**, 187 (1997).
18. L. Hua, H. Rafii-Tabar, M. Cross: Philos. Mag. Lett **75**, 237 (1997).
19. H. Rafii-Tabar, L.Hua, M. Cross: J. Phys.: Condens. Matter **10**, 2375 (1997).
20. R.K. Pathria: *Statistical Mechanics* (Pergamon Press, Oxford, 1972).
21. S. Nosé: J. Chem. Phys. **81**, 511 (1984); W.G. Hoover: Phys. Rev. A **31**, 1695 (1985).
22. A.P. Sutton, J.B. Pethica: J. Phys.: Condens. Matter **2**, 5317 (1990).
23. M. Born, K. Huang: *Dynamical Theory of Crystal Lattices* (Clarendon Press, Oxford, 1954).
24. F.F. Abraham et al.: Phys. Rev. Lett. **73**, 272 (1994).

Simulation of Dislocation Core, Plastic Deformation and Fracture in Metallic Crystals

Masao Doyama

Teikyo University of Science & Technology, Uenohara, Yamanashi 409-0133, JAPAN

Summary. The core structure and core energy were calculated for complete edge dislocations and dissociated edge dislocations in copper. By making a step on one surface (102) of a rectangular small paralellpiped copper crystal, dislocations could be created by the molecular dynamic method. The dislocation created was not a complete edge dislocation but a pair of Heidenreich— Shockley partial dislocations. Each time a dislocation was created, the stress on the surface was released. Small copper and iron crystals having a notch were pulled (until fracture), compressed and buckled by use of the molecular dynamic method. An embedded atom potential was used to represent the interaction between atoms. Dislocations were created near the tip of the notch. A very sharp yield stress was observed. An amorphous iron specimen was strained also extended until it fractured.

Key Words. Simulation, dislocation core, creation of dislocations, tensile deformation, buckling, copper, iron, single crystals, amorphous iron

1. Introduction

Micromachining and nanomachining are becoming popular and it is important to clarify the micromechanism of plastic deformation and fracture of small crystals. It is well known that the plastic deformation of metallic crystals occurs by the creation, motion and cutting of dislocations. It is quite important to know where the dislocations are created during the deformation and how they move in specimens when they are deformed. In this paper, the molecular dynamics method is used to simulate the deformation of small single crystals. To represent the interaction potentials between atoms, pairwise potentials are often used, but in metals especially the effect of the surface is important and the pair potentials are not ideal to represent the interactions near the surface. Embedded atom potentials have been developed to overcome this difficulty. Here, the creation and motion of dislocations in copper and iron crystals and amorphous iron are simulated by molecular dynamics with embedded atom potentials.

2. Interatomic Potentials

2.1 Pair Potentials

The parameters of truncated Morse potentials have been calculated for copper [1], silver, gold and aluminum [2]. The interaction energy between the *i*-th and *j*-th atoms is

$$\varphi(r_{ij}) = D\,[\exp\{-2\alpha\,(r_{ij}-r_o)\} - 2\exp\{-\alpha\,(r_{ij}-r_o)\}], \tag{1}$$

where $D = 0.171337\text{eV}$, $\alpha = 2.320783\,\text{Å}^{-1}$, $r_o = 2.588754\,\text{Å}$ for including the third neighbor truncation for copper [1]. The parameters were determined using the formation of a vacancy which is one of the earliest trials known [2]. The pair interaction in iron[3] was represented by

$$\varphi(r_{ij}) = a_1\,(r_{ij} - a_2)^4 + a_3\,(r_{ij} - a_4)^2 + a_5\,, \tag{2}$$

where $a_1 = -0.188917\,\text{eV}\,\text{Å}^{-1}$, $a_2 = 1.82709\,\text{Å}$, $a_3 = 1.70192\text{eV}\,\text{Å}^{-1}$, $a_4 = 2.50849\,\text{Å}$, $a_5 = 0.198294\text{eV}$.

2.2 Embedded Atom Potentials

In metals, the conduction electrons transfer from one atom to another and the interaction cannot be represented by pairwise potentials. The interaction between the *i*-th atom and the *j*-th atoms depends not only on the distance between them but also on the positions of other atoms around them. Copper was chosen as an example of a face centered cubic lattice. The n-body embedded function proposed by us [4] was used.

The total energy is given by

$$\begin{aligned} E_{\text{total}} &= \Sigma\, E_i \\ E_i &= F(\rho_i) + (1/2)\,\Sigma\,\Phi(r_{ij}) \\ r_{ij} &= |\,r_i - r_j\,| \\ \rho_i &= \Sigma\, f(r_{ij}) \end{aligned} \tag{3}$$

Oh and Johnson [5] used

$$F(\rho) = a\,(\rho/\rho_e)^n + b\,(\rho/\rho_e)\,. \tag{4}$$

Here E_{total} is the total internal energy, E_i is the internal energy associated with atom i. ρ_i is the electron density at atom i due to all other atoms, $F(\rho_i)$ is the embedding energy of atom into electron density ρ_i. $\Phi(r_{ij})$ is the two body central potential between atoms i and j separated by r_{ij}. $f(r_{ij})$ is the contribution to the electron density at atom i due to atom j at the distance r_{ij} from atom i.

$$f(r) = f_{old}(t) - f_c(r)$$
$$f_{old} = f_e \exp\{-\beta(r/r_e) - 1\}$$
$$f_e(r) = f_{old}(r_c) + g(r) f'_{old}(r_c) / g'(r)$$
$$\Phi(r) = \Phi_{old}(r) - \Phi_c(r) \tag{5}$$
$$\Phi_{old}(r) = \Phi_e \exp\{-\gamma(r/r_e - 1)\}$$
$$\Phi_c(r) = \Phi_{old}(r_c) + g(r)\, \Phi'_{old}(r_c) / g'(r_c)$$
$$g(r) = 1 - \exp\{\delta(r/r_e) - (r_c/r_e)\}$$

For copper Oh and Johnson [5] give $\beta = 5$, $\gamma = 8.5$, $\delta = 20$, $r_c = 1.9\, r_e$, $\phi_e = 0.36952\text{eV}$, $a = -4.0956$, $b = -1.6979$, $n = 0.44217$, and $\rho_e =$ 12.793.

We [4] have used

$$\Phi(r_{ij}) = a_1 (r_{c1} - r_{ij})^2 \exp(-c_1 r_{ij}) \tag{6}$$
$$f(r_{ij}) = a_2 (r_{c2} - r_{ij})^2 \exp(-c_2 r_{ij})$$

where $a_1 = 8289.46000$, $a_2 = 0.0183251035$, $c_1 = 10.727291$, $c_2 = 0.319759370$ and $d = 13.0792125$ for copper. In this case the unit is the nearest neighbor distance.

3. Core of Dislocations in Copper [6]

The elastic-continuum treatment of dislocations in metal crystals has always been suffered from difficulties associated with the dislocation core. The strain near the core is so severe that the displacements and energies near the core cannot be treated by standard elastic theory. It was said that the core was hollow or molten. By continuum elastic theory, the energy within radius r from the center of a dislocation can be expressed by

$$E = (\mu b^2/(4\pi)) \ln (r/r_0). \tag{7}$$

By the elastic theory, the energy outside r_0 can be treated and the energy inside r_0 cannot be determined. In other words, the core energy cannot be determined by the elastic theory. Using molecular dynamics or molecular statics, the value of r_0 can be determined. Figure 1 shows the E verusus R relation for (a) elasto-atomic, in which the atomic positions are given by the eleastic theory, (b) complete edge dislocation and (c) dissociated edge dislocation. *It* was found that no activation energies were required to split complete dislocations to dissociated dislocations in copper, i.e. the complete dislocations split into Heidenreich - Schockley partial dislocations without activation energies. Figure 2 shows the projection of positions of the atoms above the slip plane (o) and below the slip plane (▲). Figure 3 shows the projection of atomic positions to the (112) plane. The core of dislocations is filled with atoms:it is not hollow or molten.

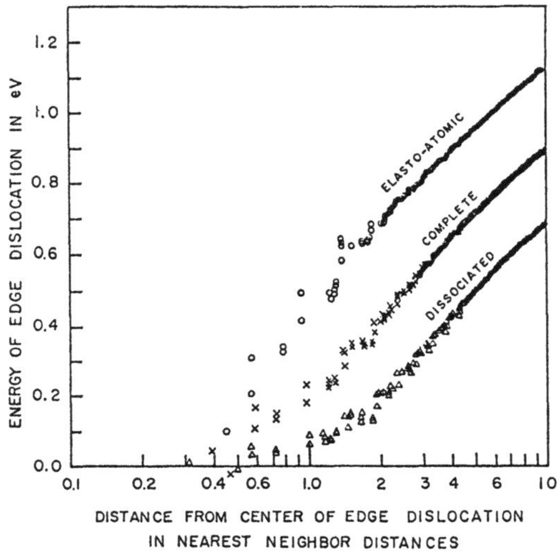

Fig. 1. Energy within a given radius as a function of that radius as measured from the center of dislocation. The energy given is for a length $d_0/[2(3)^{1/2}]$.

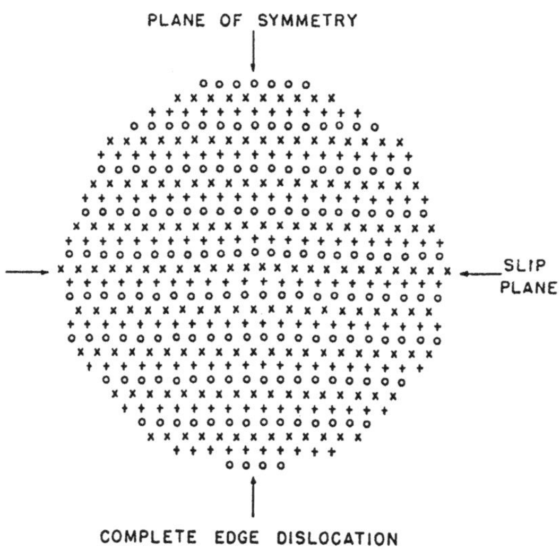

Fig. 2. Positions of atoms in three consecutive {112} planes normal to the complete edge. The core is neither hollow nor like a liquid and the distorsion in the core is not very severe.

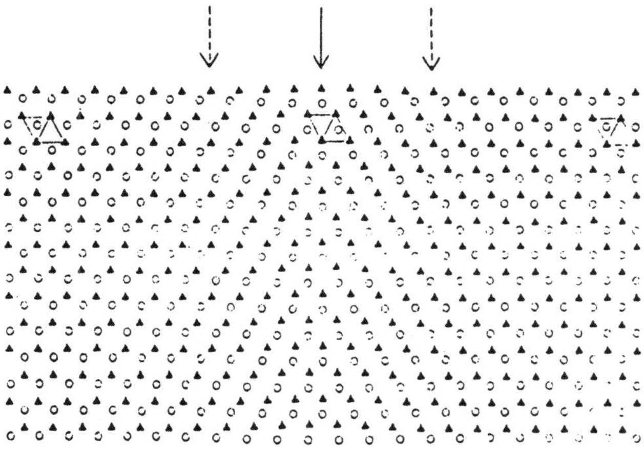

Fig. 3. Position of atoms in two (111) planes, one above (triangle) and one below the slip plane of the dissociated edge dislocation. The position of the original complete dislocation line is along the (112) direction indicated by the full arrow. The partial dislocations, indicated by the dashed arrows, and the region of a stacking fault which separates them can be easily seen by observing the figure from either side at a low angle.

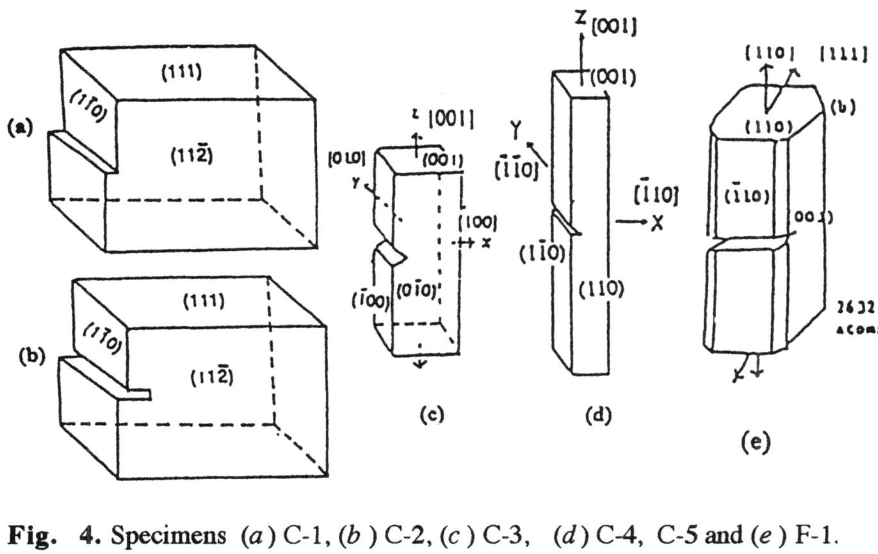

Fig. 4. Specimens (*a*) C-1, (*b*) C-2, (*c*) C-3, (*d*) C-4, C-5 and (*e*) F-1.

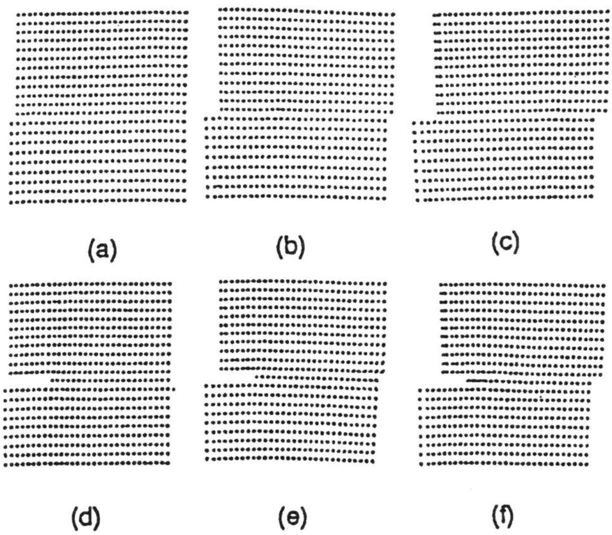

Fig. 5. Projection of copper atoms of specimens C-1 and C-2. Creation of dislocations. (**a**), (**b**), and (**c**) are specimen C-1 without a notch and (**d**), (**e**) and (**f**) are for specimen C-2 with a notch.

4. Creation of Dislocations [7]

Specimen C-1(Fig. 4a) is a rectangular parallelepiped having the faces of (111), (111), (112), (112), (110) and (110). *x*, *y*, and *z* directions were taken in the [110], [112] and [111] directions, respectively. Specimen C-2 is similar to Specimen C-1 but with a notch on the (110). A step (height = 0.01 , where *d* is the nearest neighbor distance) was made on the (110), then all the atoms in the crystal were relaxed 50 cycles by molecular dynamics. The step height was again increased by 0.01d. This process was repeated.

A Heidenreich - Schockley (H - S) partial dislocation was created followed by the other H - S partial dislocation. As soon as the first dislocation was created the stress was greatly reduced; a very sharp yield point was observed. Figure 5 shows the projection of atoms on the (112) plane. Figure 6 shows the projection of atoms two layers above the slip plane and one layer below the slip plane on the (111) plane. Stacking faults are clearly seen. The stress—strain curves for Specimen C-1 and C-2 are plotted in Fig. 7. Each time the half-dislocation was introduced, a sharp yield stress was observed. Two half-dislocations are connected with the stacking fault.

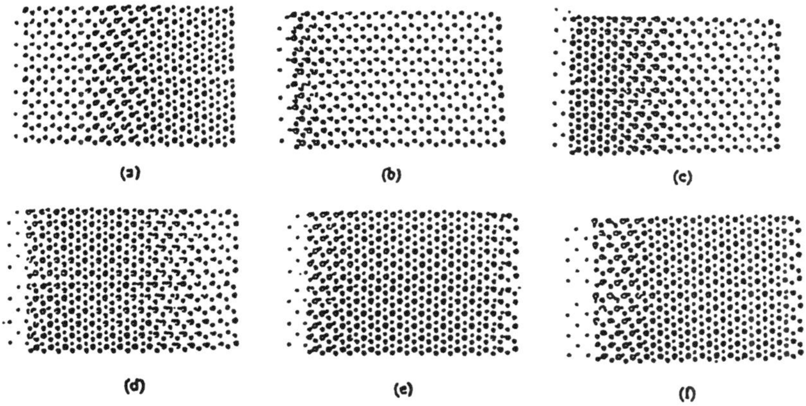

Fig. 6. Projection of atoms in two layers above the slip plane and one layer below the slip plane on the (111) in specimen C-1. Two half-dislocations are connected with a stacking fault. (*a*) 3500 time steps (0.7 *d* step height), (*b*) at 5000 time steps (1.0 *d* step height), (*c*) at 6000 time steps (1.2 *d* step height), (d) 7,000 time steps (1.4 *d* step height), (*e*) at 7500 time steps (1.5 *d* step height), and (*f*) 9500 time steps (1.9 d step height).

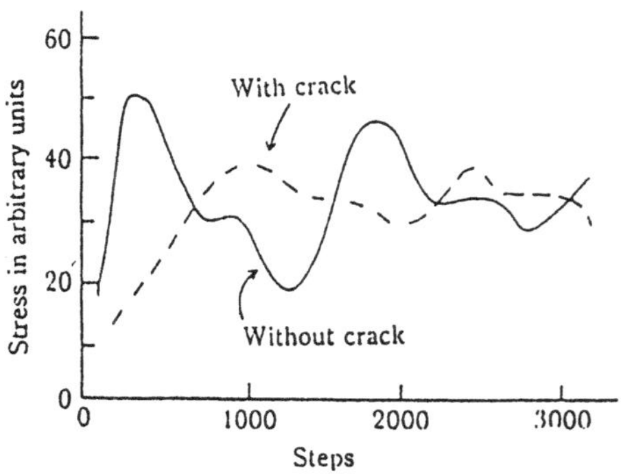

Fig. 7. Stress-strain curves for specimens C-1 and C-2.

5. Deformation and Fracture of Single Crystals [8]

Specimen C-3 (Fig. 4c) contains 4338 atoms having the faces of (100), (100), (010), (010), (001), and (001). the x, y, and z directions were taken to be in the direction of [100], [010], and [001], respectively. The size of specimen C-3 is $8a \times 8a \times 26a$, where a is the lattice parameter and a notch was made in the y direction at the center on the (100). The surfaces were free and the periodic boundary conditions were not used. Specimen C-3 was pulled in the [001] direction by 0.5%, i.e. all the atoms were deformed uniformly, then all atoms except the one layer of each end were relaxed 25 cycles, and again were pulled 0.5% in the [001] direction, and all atoms were relaxed 25 cycles. The atoms in the end layers were relaxed only in x and y directions not in z direction. This procedure was repeated. The time step of the molecular dynamics was 5×10^{-15} s.

Atomic positions near the center of Specimen C-3 projected on the xz plane (010) are plotted in Figs. 8a–d. In Fig. 8b, near the tip of the crack, a half-dislocation was created. Two other half-dislocations are started near the grip of Specimen C-3. These can be clearly seen when the reader looks at the figures at a small angle to the paper. In Figure 9 the stress–strain curve is plotted. A very sharp yield point was found.

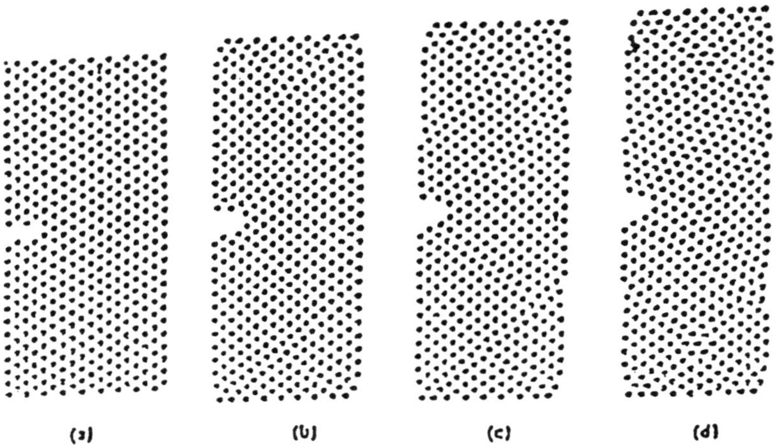

Fig. 8. Specimen C-3 (a) 300 time steps (6% elongation), (b) 375 time steps (7.5% elongation), (c) 500 time steps (10% elongation), and (d) 675 time steps (13.5%). All are projected on xz (010) plane.

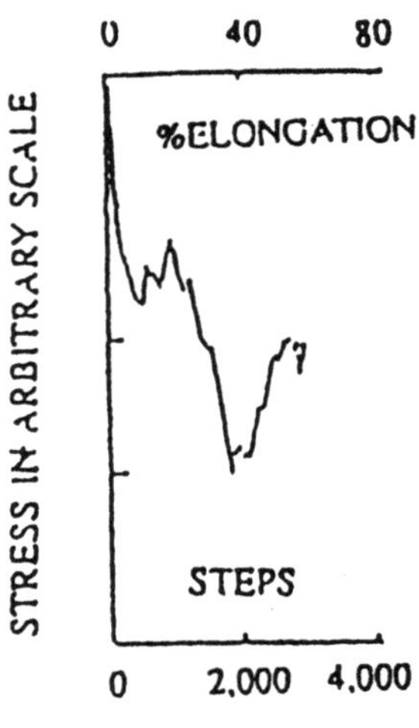

Fig. 9. Stress–strain. curve for specimen C-3.

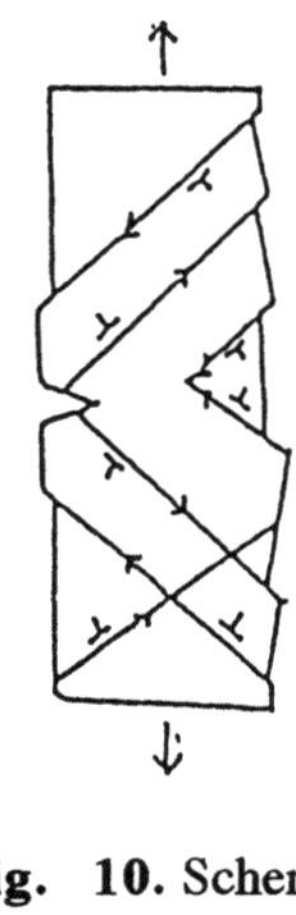

Fig. 10. Schematic diagram of creation and motion of half dislocations in specimen C-3.

Fig. 11. Atomic position just before. fracture.

It was found that a notch is normally a source of dislocations due to the stress concentration. [211] (111) and [121] (111) half-dislocations were created leaving stacking faults. A schematic diagram is given in Fig. 10 for specimen C-3. The initial stage of deformation was clearly conducted by the creation and motion of dislocations. In the middle stage, dislocations on different slip planes cross each other. The work hardening occurs in this stage. In the final stage, in the case of copper, a crystalline state with many dislocations occurs and the inhomogeneous deformation is normal. Figure 11 shows the atomic positions just after fracture. A polycrystalline state near the fracture was observed.

6. Dynamics of Buckling [9]

Specimen C-4 (Fig. 4d) is a rectangular parallelepiped having the faces of (110), (110), (110), (110), (001) and (001). A notch was introduced near the center of (110). The x-axis was taken to be [110], y [110], and z [001]. The size of specimen C-4 is 5d $\times$ 5d $\times$ 50a, where d is the nearest neighbor distance and a is the lattice parameter. Specimen C-4 contains 2540 atoms. Specimen C-4 is a perfect crystal with a notch.

Buckling is an important plastic deformation of metallic materials. Specimen C-4 was uniformly compressed in the [001] direction and at the same time all the atoms near the center of the z-axis (one half of the specimen near the center of the z direction) were displaced in the x direction sinusoidally and all the atoms were relaxed. After every 150 cycles this deformation was given. After giving a small compression and sinusoidal displacement, all the atoms in the specimen were relaxed for 150 time cycles and then deformed again. This process was repeated. After 5000 time cycles, only compression was given to all the atoms.

The time step of the molecular dynamics was 3×10^{-15}s.

The projections of atom positions of Specimen C-4 on the xz-plane (110) are given in Fig. 12. Uniform deformation occurred throughout the specimen except very close to the notch. Above 5000 time cycles, the displacements in the x direction were not given but only uniform compression in the z direction was given. In Fig. 12b at 5000 time cycles some irregularity can be seen near the one third of the specimen in the z direction from both ends. These regions became small crystals, as seen in Fig. 12f and g. Near the center of the specimen a few crystallites were nucleated.

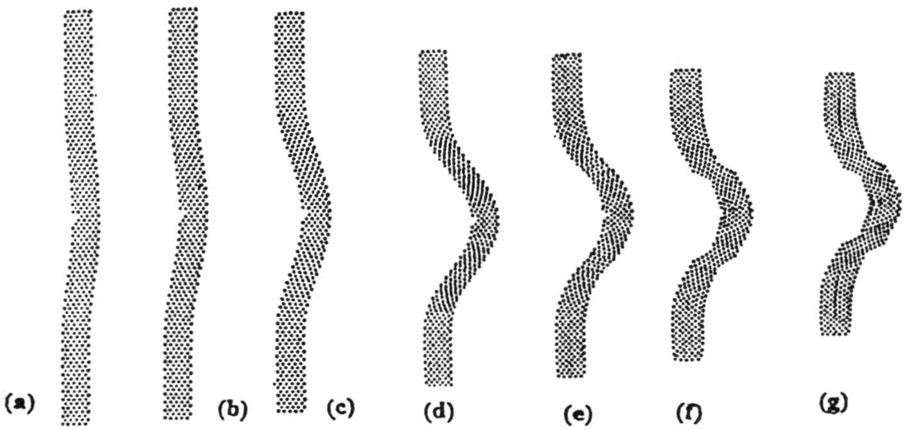

Fig. 12. The projections of atom positions of Specimen C-4 on the *xz*-plane, (110), (a) at 500 time steps, (b) 1000 time steps and (c) at 1500 time steps. (d) 5000 time steps, (e) 7500 time steps, (f) 125000 time steps, and (g) 17500 time steps.

7. Compression [10]

Specimen C-5 is also a rectangular parallelepiped having the faces of (110), (110), (110), (110), (001) and (001) and contans 4896 atoms. The size of Specimen C-6 is $12d \times 8a \times 20d$. Specimen C-6 has a notch at the center of the (110).

Specimen C-5 was uniformly compressed. Dislocations started near the tip of the notch above 2000 time cycles (Fig. 13b). Bulging can be observed above 3500 time cycles (Fig. 13c). Figur 13d is at 7500 time cycles, Fig. 13e at 9000 time cycles and Fig. 13f is at 15000 cycles. Severe bulging is observed.

When no displacements in the direction perpendicular to the axis were given, no buckling was observed but only compression and bulging was observed. When displacements in the direction perpendicular to the axis were given, buckling started. In this case, a single crystal became a polycrystal.

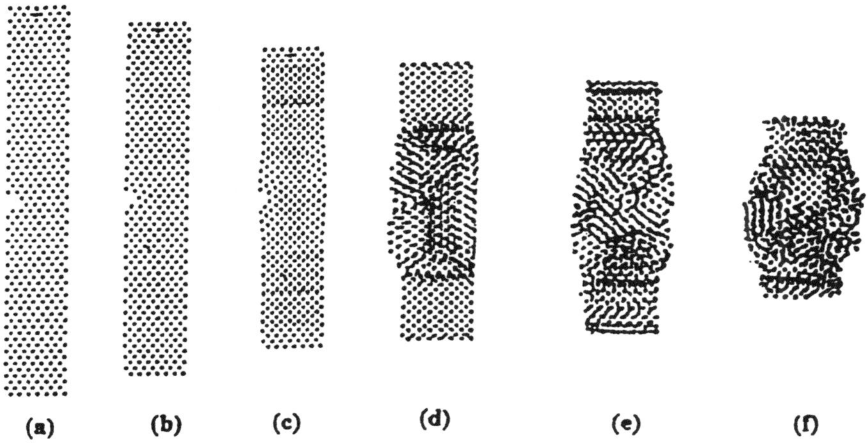

Fig. 13. The projection of atom positions of Specimen C-5 on the xz-plane (110). (a) at 500 time steps, (b) 2000 time steps, (c) 4500 time steps, (d) 7500 time steps, (e) 9000 time steps, and (f) 15000 time steps.

8. Plastic Deformation and Fracture of Iron Whiskers [11]

Specimen F-1 (Fig. 4e) is a needle shaped whisker, consisting of 2886 iron atoms whose side faces are (110), (110), (001), (001), (110), and (110). The iron whisker was pulled using the method of molecular dynamics. Figure 14 shows the projection of atoms onto the (001). The stress – strain curve is plotted in Fig. 15. Deformation started near the edge of a crack. Necking was observed and finally the specimen was completely broken into two pieces. A very sharp yield stress was found (Fig. 15).

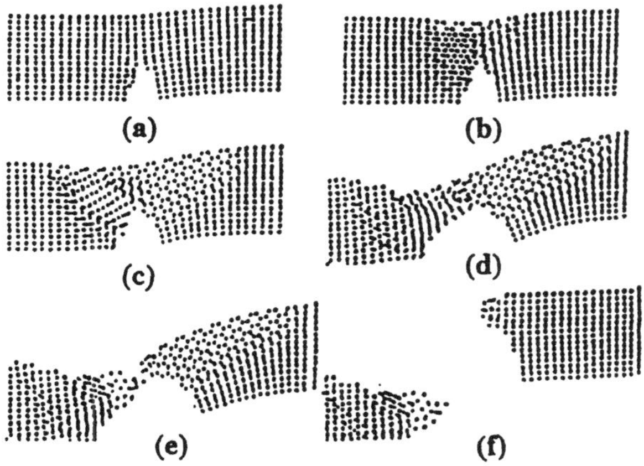

Fig. 14. Projection of atom positions of specimen F-1 on (100).

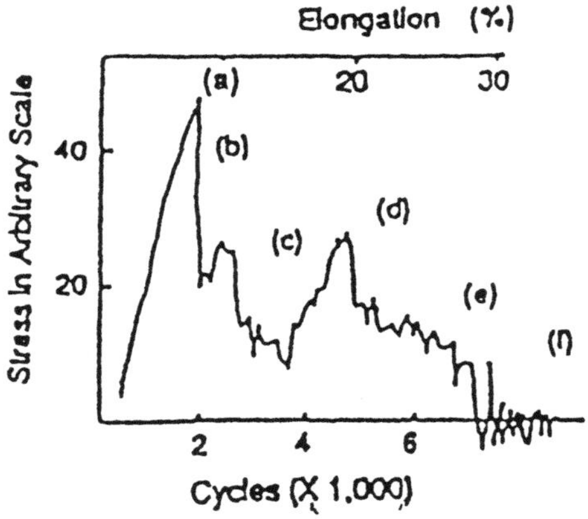

Fig. 15. Stress—strain curve for iron specimen F-1. (a), (b), (c), etc. correspond Fig. 14a, b, c, etc.

9. Plastic Deformation of Amorphous Iron [12]

An amorphous iron cylinder is represented in specimen A-1 (Fig. 16). The cylinder was pulled by the method of molecular dynamics. The internal energy increased linearly with elongation and suddenly dropped (Fig. 17). The linear part corresponds to an elastic deformation (Fig. 16a) and the sudden decreasing part corresponds to the plastic deformation (Fig. 18b). Atoms below a certain plane which is nearly 45° from the extension axis moved below and atoms above the plane moved in the opposite direction.

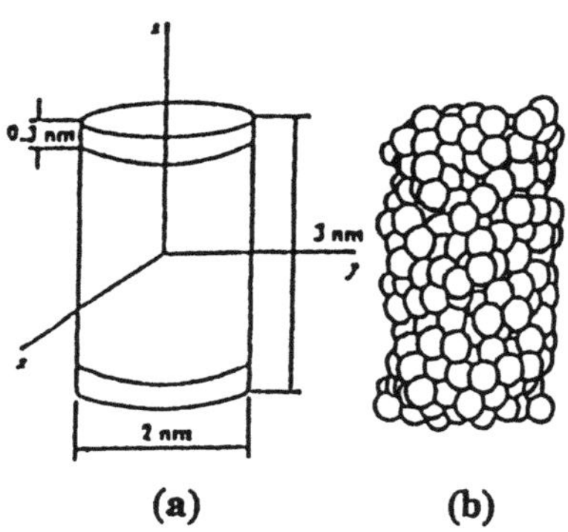

Fig. 16. Cylindrical amorphous specimen A-1.

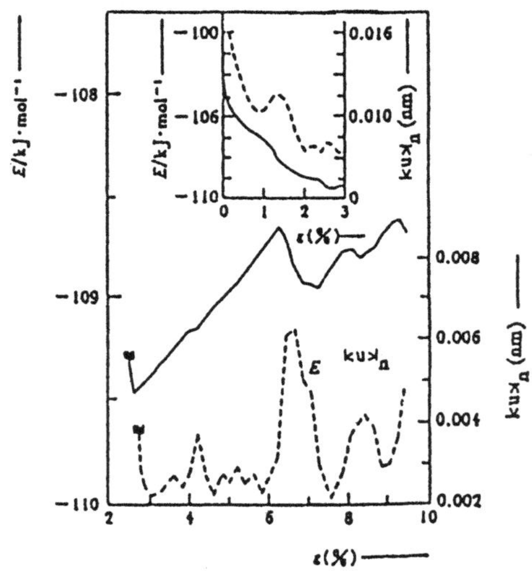

Fig. 17. Molar average mean energy per iron atom and mean atomic displacement during extension.

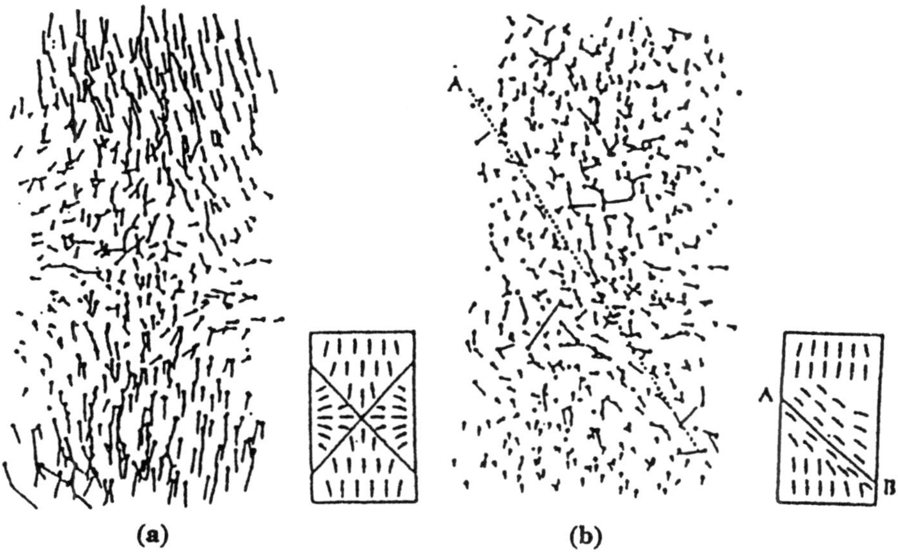

Fig. 18. Projection of amorphous iron atoms on xz plane, (a) nominal elongation from 2.6 to 6.2%, (b) nominal elongation from 6.2 to 7.5%.

10. Conclusion

Plastic deformation and fracture in metals can be understood by use of molecular dynamics. Some of the general features do not depend upon the details of the effective potentials. Details of plastic deformation and fracture are not very well known.

ACKNOWLEGEMENT

This research is supported by the Grant-in-Aid by the Ministry of Education, Science and Culture under Priority Area "Physics and Chemistry of Functional Graded Materials".

References

1. M. Doyama, R.M.J. Cotterill: Phys. Rev. **137**, A994−999 (1964) .
2. R.M.J.Cotterill and M. Doyama: *Lattice Defects and their Interactions,* Ed. R.R. Hasiguti, (Gordon Breach, New York, (1967), 1−78, M. Doyama, R.M.J. Cotterill, ibid pp. 80−185.
3. H.M.Pak, M. Doyama: Journal of the Faculty of Engineering, University of Tokyo, **(B) 30**, 111 (1969).
4. M. Doyama, Y. Kogure: Radiat. Eff. Defects Solids, **142**, 565−572 (1997).
5. D.J. Oh, R.A.Johnson: J. Mater. Res. **3**, 471 (1986). D.J. Oh and R.A. Johnson: "*Atomistic Simulation of Materials*", Eds V. Vitek, D.J. Srolovitz, (Plenum Press, New York, 1989) pp.233.
6. R.M.J. Cotterill, M. Doyama: Phys. Rev. **145**, 465−478 (1966) . M. Doyama, R.M.J. Cotterill, Phys. Rev. **150**, 448−455 (1966).
7. M. Doyama: (to be published).
8. M. Doyama: Nanostruct. Mater. **9**, 689−692 (1997)
9. M. Doyama: (to be published).
10. M. Doyama (to be published).
11. Masao Doyama, R. Yamamoto, *Macro- and Micro-Mechanics of High Velocity Deformation and Fracture*, eds K.Kawada, J. Shioiri, (Springer-Verlag Berlin, 1987). Masao Doyama, Ryoichi Yamamoto, *Dislocations in Solids,* Eds H. Suzuki, T. Ninomiya, K. Sumino and S. Takeuchi, (The University of Tokyo Press, 1985), pp.85−88.
12. R. Yamamoto, H. Matsuoka, M. Doyama: Phys. Status Solidi A **51**, 163 (1979).

Molecular Dynamics Analysis of Nanometric Metal Cutting Mechanism

Shoichi Shimada[1] and Naoya Ikawa[2]

[1]Department of Precision Science and Technology, Osaka University, 2-1 Yamada-oka, Suita, Osaka 565-0871 Japan
[2]Department of Intelligent Machine Engineering, Osaka Electro-Communication University, 18-8 Hatsu-cho, Neyagawa, Osaka 572-8530 Japan

Summary. A review is made of the present situation of molecular dynamics simulation of the nanometric metal cutting mechanism. Chip formation and surface generation mechanisms are well clarified by the simulations. The ultimate accuracy attainable is also predicted in terms of the behavior of the atoms in the workpiece. Cutting temperature, as calculated in the simulation, can be useful for prediction and control of tool wear. Some phenomena are also predicted in ultra high speed cutting which are not achieved experimentally as yet. Finally, problems and limitations of the simulation are discussed.

Key words. ultraprecision metal cutting, diamond turning, computer simulation, molecular dynamics, cutting mechanism, machining accuracy, cutting temperature, high speed cutting

1. Introduction

Ultraprecision metal cutting technique, which is called single point diamond turning (SPDT) in some cases, has been playing very important roles in manufacturing of a variety of key optical, electronic and mechanical components in the systems of advanced science and technology [1]. The principle of metal cutting is surface generation due to the material removal by a cutting tool from a workpiece as shown in Fig. 1. The relative motion for material removal between the cutting tool and workpiece is provided by a machine tool on which both of them are mounted. Therefore, the technique basically relies on an extremely high fidelity in the transfer characteristics of both machine tool motions and cutting edge profile into the work surface under precisely controlled machine and environmental conditions. Surface roughness and form accuracy of the products have been attained on the order of 1 nm and 100 nm/100 mm, respectively.

The machine tool, which is the key element for this technique, has attained an extremely high level in its performance through scientific and engineering efforts. The accuracy of such machine tools have reached the order of 10 nm on practical machines and of 1 nm or less on experimental ones [2-4]. With the use of a specially prepared fine diamond cutting tool on a highly reliable machine tool, it has been experimentally confirmed that very fine chips, the undeformed thickness

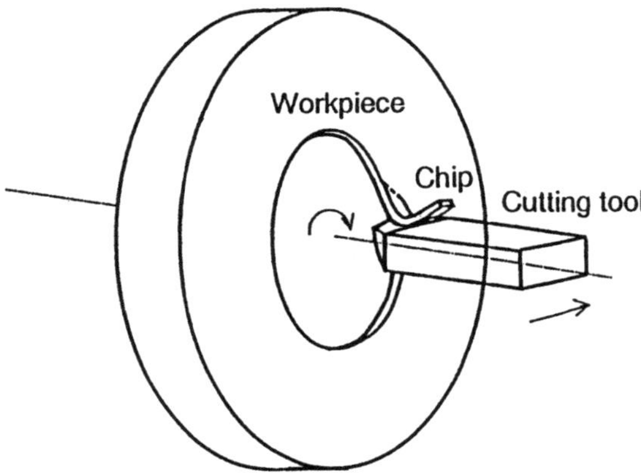

Fig. 1. Principle of metal cutting (face turning).

of which is down to 1 nm, can be stably removed in diamond turning of free machining work materials [5].

However, the ultimate accuracy attainable in microcutting and details of microcutting phenomena, such as chip formation and surface generation mechanisms, are not well understood, because of the inevitable difficulties in experimental evaluation and/or analysis due to limitation in controllability of machine motion and in observation and measuring techniques. Furthermore, the microcutting phenomena which take place in a small region can be atomistic or discrete in nature rather than continuous, as assumed in conventional continuum mechanics. For the analyses of these atomistic processes, which cannot be confirmed experimentally, molecular dynamics (MD) computer simulation is a useful tool. Based on the simulation, several kinds of analyses of microcutting processes have been developed [6-14].

In this paper, several potential attempts at a molecular dynamics approach for understanding the cutting process and for determining the ultimate accuracy attainable in ultraprecision metal cutting have been reviewed.

2. Molecular Dynamics Simulation of Nanometric Cutting Phenomena

MD simulation is a useful method to analyze accurately the atomistic behavior of deformation and/or fracture processes of an atomic lattice model. Using the simulation, progressive motion, including lattice vibration of individual atoms which compose the model, can be traced by solving sequentially the differential equations of motion of all the atoms. The behavior of the solid model can be analyzed from the synthesized motion of individual atoms.

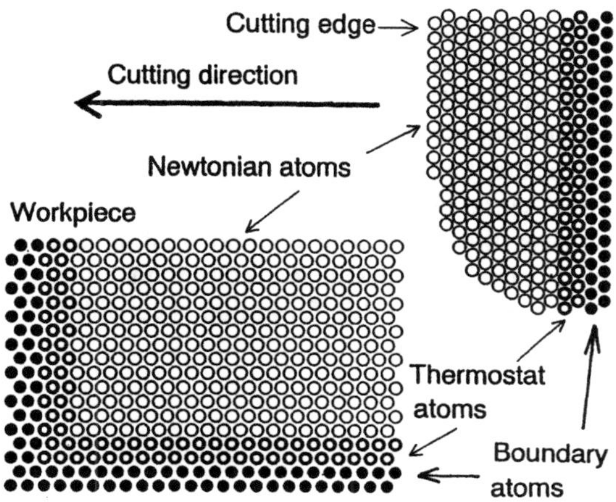

Fig. 2. Schematic model of microcutting for molecular dynamics simulation.

To analyze the nanometric cutting process, the MD simulations are carried out assuming two-dimensional orthogonal cutting of the (111) plane of hypothetical Morse-potential type copper and aluminum cut by the (111) plane of a diamond cutting edge, as shown in Fig. 2 [9,11]. The static boundary atoms are fixed to the rigid bases of the workpiece and cutting edge. The atoms in two layers adjacent to the boundary atoms act as thermostat atoms. In these layers, the velocities of atoms are adjusted for every specific time step of computation so as to stabilize the average temperature of thermostat layers at room temperature (293 K). For the computation time step, 1 fs and 10 fs are employed depending on the purpose of the analysis. The simulations of microcutting are carried out mainly under the cutting speed of 20 m s^{-1}. The cutting speed of 200 m s^{-1} is also employed purposely to reduce computation time.

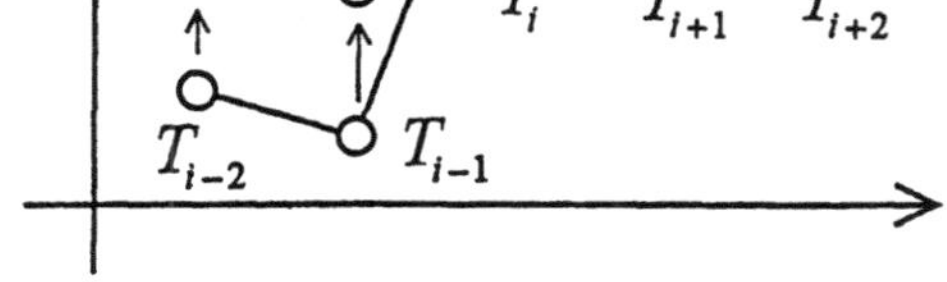

$$T_i \rightarrow T_i'$$

$$T_i' = T_i - \frac{(T_i - T_m)\beta}{k_B}$$

$$T_m = \frac{1}{N}\sum_{i=1}^{N} T_i$$

Fig. 3. Scaling of equivalent temperature of atoms to accelerate the thermal conduction and consequently to reduce the thermal field gradient in the model. The coefficient β is so chosen that the simulated thermal conductivity coincides with the macroscopic value measured so far.

In general, thermal conduction in metals cannot be analyzed correctly in MD simulation, because electron behavior is not taken into account in conventional MD models. As thermal conduction in metals is mainly governed by electron mobility, metals for practical use show considerably higher thermal conductivity, and consequently a smaller gradient in the thermal field, than those estimated by MD simulation in terms of lattice vibration. Therefore, for more realistic simulation of the microcutting process, the gradient in thermal field should be scaled so as to coincide with that calculated using the continuum thermal conduction theory [14], as shown in Fig. 3, because chip formation is considerably affected by the thermal conditions in the shear zone.

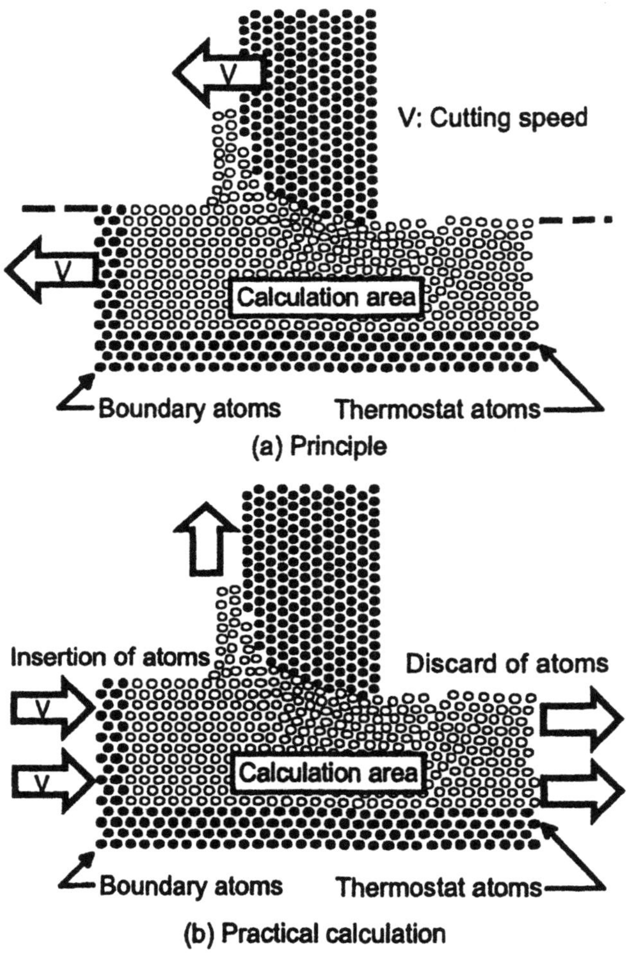

Fig. 4. Scheme of translational calculation area method.

For the analysis of cutting phenomena under steady-state chip removal, the translational calculation area method, by which the cutting distance can be infinitely extended, is used in the simulations [15]. To limit the calculation area in which the behaviors of individual atoms are analyzed by the MD method, boundary and thermostat layers move at the cutting speed with the cutting edge, as shown in Fig. 4a. In practical calculations, new atoms are successively inserted from the left into the calculation area, in which the cutting edge is fixed, and the atoms that leave the calculation area are thrown away from the right or top as shown in Fig. 4b.

3. Atomistic Processes of Chip Formation and Surface Generation

Figure 5 shows an example of a MD simulation for the microcutting of copper [11]. The cutting force per unit width of the cutting edge obtained from MD simulations and cutting experiments are summarized in Fig. 6 as a function of the nominal undeformed chip thickness [11]. Comparative results show that a fairly reasonable correlation between the chip morphologies and the size effects in the cutting force is obtained from both simulations and experiments. The results suggest that MD simulation can be a useful tool for the analysis of nanometric cutting phenomena.

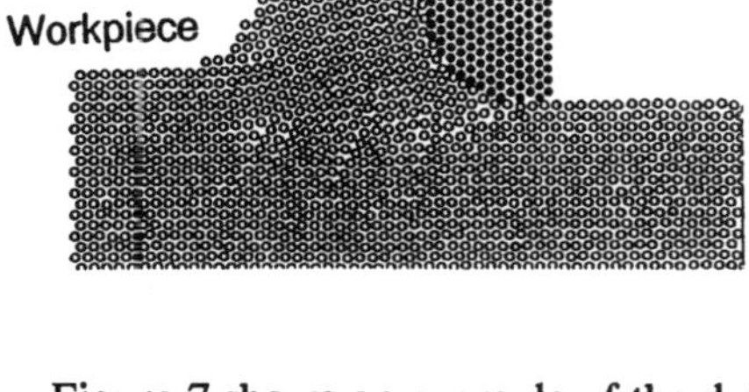

Fig. 5. Example of MD simulation in copper microcutting (cutting speed : 20 m s^{-1}, undefromed chip thickness : 0.5 nm, cutting edge radius : 2 nm).

Figure 7 shows an example of the deformation behavior of the workpiece atoms in front of the cutting edge. Due to ploughing of the cutting edge, the lattice of the workpiece is deformed as if by buckling. When the strain energy stored in the deformed lattice exceeds a specific level, the atoms begin to rearrange so as to relax the lattice strain. However, since the strain energy stored is not sufficient for perfect rearrangement, some dislocations are generated in the workpiece lattice. With progress of the cutting edge, many dislocations are generated successively in the workpiece near the tool-workpiece interface in a similar manner to that

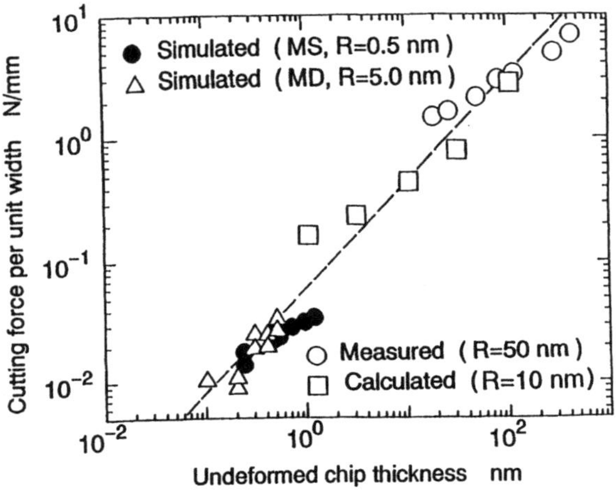

Fig. 6. Size effect in cutting force per unit width as a function of undeformed chip thickness.

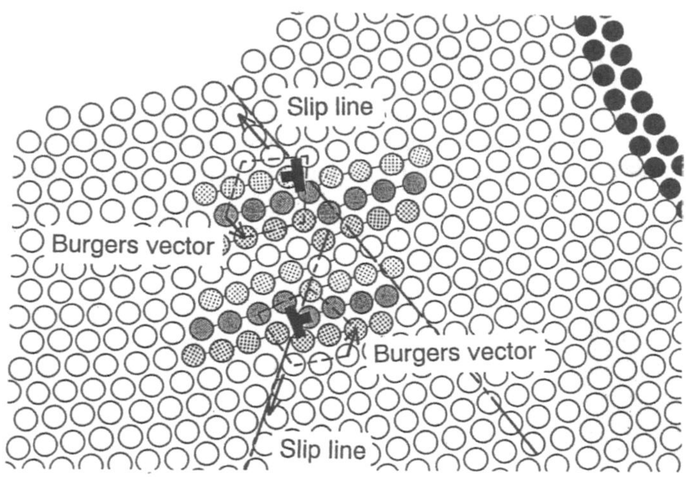

Fig. 7. Generation of dislocations in workpiece due to ploughing of cutting edge.

described above. Some dislocations move into the shear zone and disappear from the free surface. This phenomenon corresponds to an elemental process of chip formation morphologically. As a result of successive generation and disappearance of the dislocations, the chip seems to be removed stably. The other dislocations penetrate into the workpiece beneath the cutting edge. After passage of the cutting edge, all penetrated dislocations begin to move back and finally disappear from the work surface due to relaxation as a consequence of spring-back of the workpiece. As a result of the relaxation, atomistic steps are formed on the work surface. The height of the steps remaining on the work surface can be considered as the ultimate surface roughness attainable in microcutting [12].

4. Ultimate Accuracy Attainable in Microcutting

Surface finish in single point cutting by a chevron nosed tool is schematically shown in Fig. 8. Apart from machine tool performance, there are two primary factors affecting the ultimate accuracy attainable in microcutting. One is the minimum thickness of cut (MTC) which is the minimum undeformed chip thickness stably removed from a workpiece at a cutting edge under a specific cutting condition. The other is transfer fidelity of the cutting edge profile to the work surface [16,17]. The transfer fidelity may be at the same level as that of the ultimate surface roughness remaining on the work surface cut with perfect machine motion.

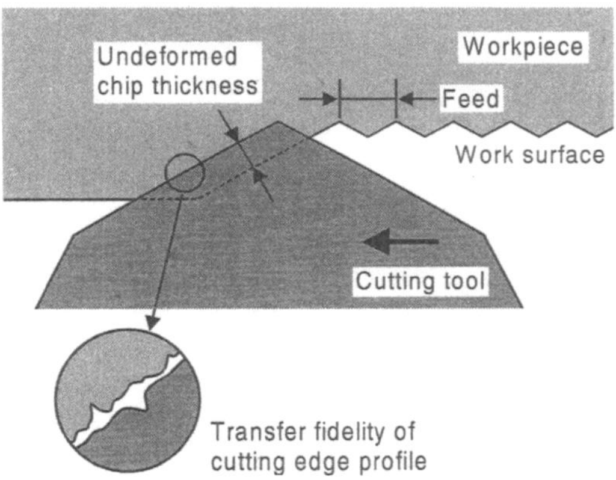

Fig. 8. Factors affecting untimate accuracy attainable in microcutting.

Figure 9 shows an example of chip formation in microcutting of copper by a cutting edge with a radius of 5 nm. While chip formation cannot be seen at an undeformed chip thickness, t_n, smaller than 0.2 nm, the initial stage of chip formation is observed at an undeformed chip thickness greater than 0.3 nm. Therefore, MTC in this cutting condition can be estimated to be 0.3 nm. Results of MD simulations with a variety of edge radii show that MTC in microcutting of copper is 1/20 to 1/10 of the radius of the cutting edge, that is, about 1 nm for a realistic fine diamond cutting tool. In cutting of aluminum, a larger MTC, which is 1/10 to 1/5 of the radius of the cutting edge, is estimated [10].

Figure 10 shows the surface roughness and deformed layer on the work surface of copper and aluminum cut by a diamond cutting edge with a radius of 10 nm. From the simulations, the transfer fidelity of the cutting edge profile to the work surface in microcutting of copper and aluminum can be expected to be 0.5 to 1.0 nm and 1.5 to 2.0 nm, respectively [10,12]. The work surface of copper seems to be free from residual strain. On the other hand, vacancies and many atoms which show high potential energy are observed on the aluminum work surface [13]. Therefore, the surface quality of aluminum is worse than that of copper in terms of both surface roughness and residual strain.

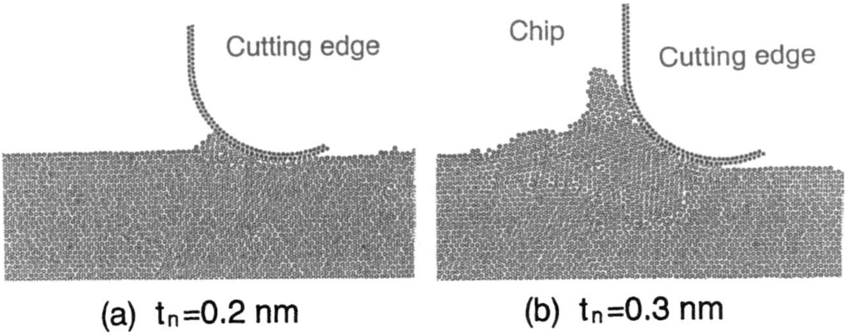

Fig. 9. Chip formation in copper microcutting (cutting speed : 200 m s^{-1}, cutting edge radius : 5 nm).

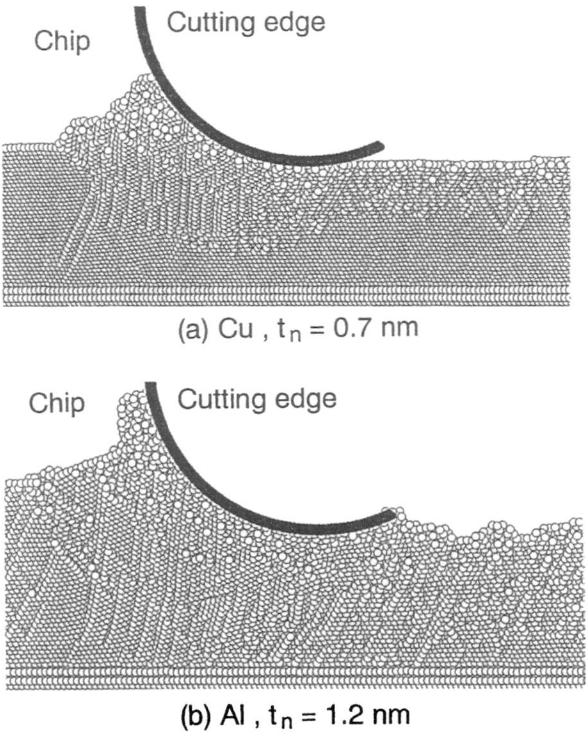

Fig. 10. Surface roughness and deformed layer on work surface in microcutting of copper and aluminum (cutting speed : 200 m s^{-1}, cutting edge radius : 10 nm, undeformed chip thickness : 0.7 nm (Cu), 1.2 nm (Al)).

MTC and transfer fidelity are intuitively considered to be affected by the affinity of the tool-work interface and/or the indentation hardness of the work material. Aluminum shows higher affinity with diamond and lower indentation hardness than copper. The effect of affinity of tool-work interface can be analyzed by changing the parameters specific to cohesive energy for the tool-work interface in the Morse potential function [5,10]. The MD simulations show that a work material which has higher affinity of tool-work interface tends to show easier chip formation and larger depth of the deformed layer. However, no remarkable difference can be seen in chip behavior and MTC. These results suggest that smaller MTC and higher transfer fidelity in microcutting are expected on a work material which shows higher indentation hardness.

The effect of grain boundaries, which are intrinsically included in practical work materials machined by diamond turning, is analyzed by the microcutting simulation using a two-grain model of copper. The simulation suggests that a nanometric distorted layer inevitably remains on a work surface of polycrystalline metals due to the dislocation trapping effect of grain boundaries [13,14].

5. Cutting Temperature and Tool Wear

Figure 11 shows the difference in the equivalent temperature of atoms converted from kinetic energy between the scaled and non-scaled MD simulations [14]. The results show that the chip formation process is significantly affected by the thermal conductivity of the work material. It is also shown that some difficulties in MD simulation for cutting temperature in metal microcutting can be eased with the introduction of a suitable scaling in the thermal conductivity. The average temperature of atoms in the shear zone calculated from the simulation is 300 to 310 K using the proposed scaling. This value is considerably lower than that (470 to 670 K) measured in the practical cutting experiments. However, uncut chip thickness in the cutting experiments are 2 to 100 μm, which are 2000 to 100 000 times larger than that in the simulation. Therefore, the average cutting temperature as calculated in nanometric cutting may not rise as high as those in the cutting experiments. On the other hand, several atoms, the equivalent temperature of which exceeds 1000 K, are observed in the chip. The atoms at this temperature can supply the activation energy necessary for chemical interaction with oxygen in the atmosphere. This result supports the observation that the surface of chips removed under nanometric uncut chip thickness is severely oxidized. The wear mechanism of the cutting tool may also be understood from an atomistic point of view based on the MD analysis of cutting temperature described above.

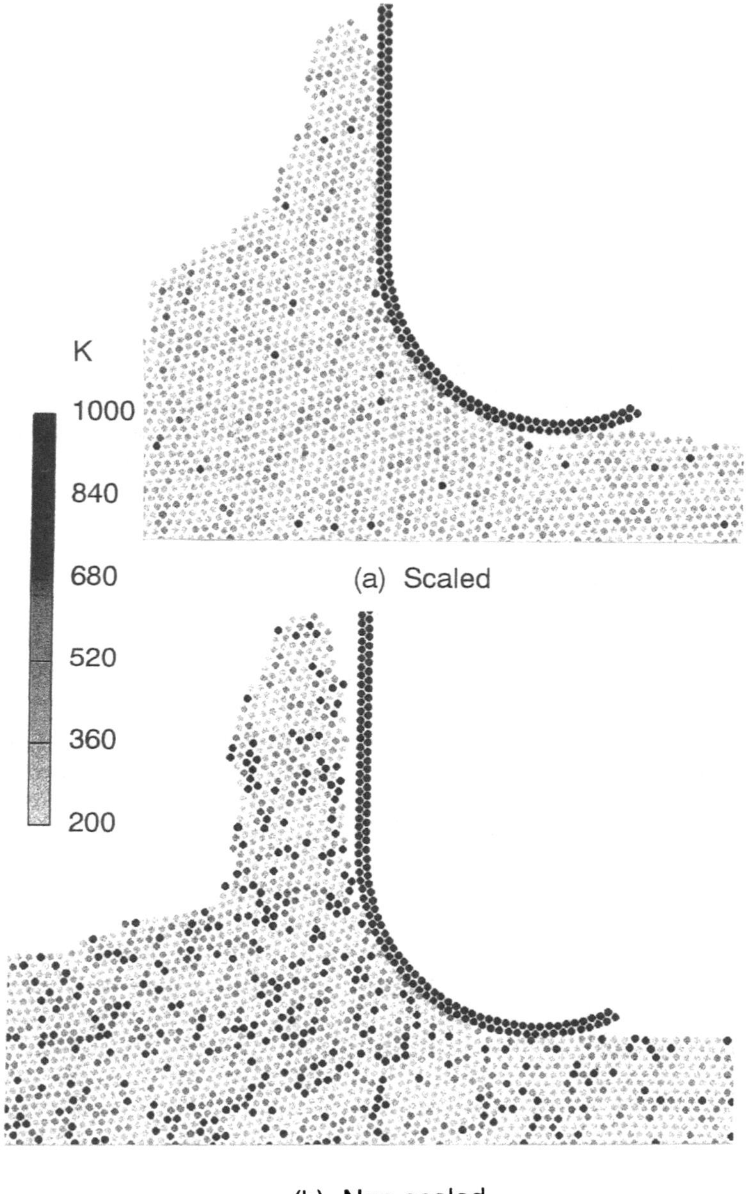

Fig. 11. Effect of scaling on distribution of equivalent cutting temperature and chip morphology (cutting speed : 20 m s^{-1}, cutting edge radius : 5 nm, undeformed chip thickness : 1nn).

6. "Hypothetical" Ultra High Speed Machining

Compatibility of high precision and productivity in nanomachining may be achieved by employing an ultrahigh cutting speed. There would be many obstacles to overcome in developing machine tools with ultrahigh levels in cutting speed and accuracy. However, it may be worthwhile to discuss the possible phenomena at very high speeds using MD simulations.

While there would be optimal values in cutting speed under specific criteria, such as surface finish and cutting force, there could arise catastrophic phenomena which could not be tolerable for a controlled machining process in an extremely high speed range. The MD simulation may give us some useful suggestions on possible phenomena which take place under hypothetical cutting conditions (perfect machine motion at ultra high cutting speed) that is not yet realized by the present technology but may be in the future.

Figure 12 shows examples of MD simulation for material removal at very high speeds [18]. The analysis suggests that a smaller depth of the deformed layer on a machined surface under a certain high cutting speed is obtained probably due to an adiabatic effect (Fig. 12a). However, it also shows a catastrophic dispersion of work material (Fig. 12b) at very high cutting speed, say 2000 m s^{-1}, at which the kinetic energy supplied by engagement of the cutting edge to the individual atoms in the workpiece exceed the specific cohesive energy of the work material. These results suggest an optimal speed exists at least in terms of surface finish.

7. Concluding Remarks

Understanding the nanometric metal cutting mechanism from an atomistic point view and predicting the ultimate accuracy attainable are of essential importance for improving the ultraprecision metal cutting technique. MD simulation can be a useful tool for these purposes. However, MD simulation is a new technique and has only recently been applied to the analyses of nanometric cutting processes. To improve the reliability and usefulness of MD simulation, some problems that must be overcome are as follows.

1. Long computation time of a few days or more;
2. Limitation in the dimensions of the model due to limited memory size of computer;
3. Difficulties in determining "realistic" interatomic potential function, especially for heterogeneous combinations;
4. Difficulties in modeling defects in solids;
5. Difficulties in applying an appropriate boundary condition.

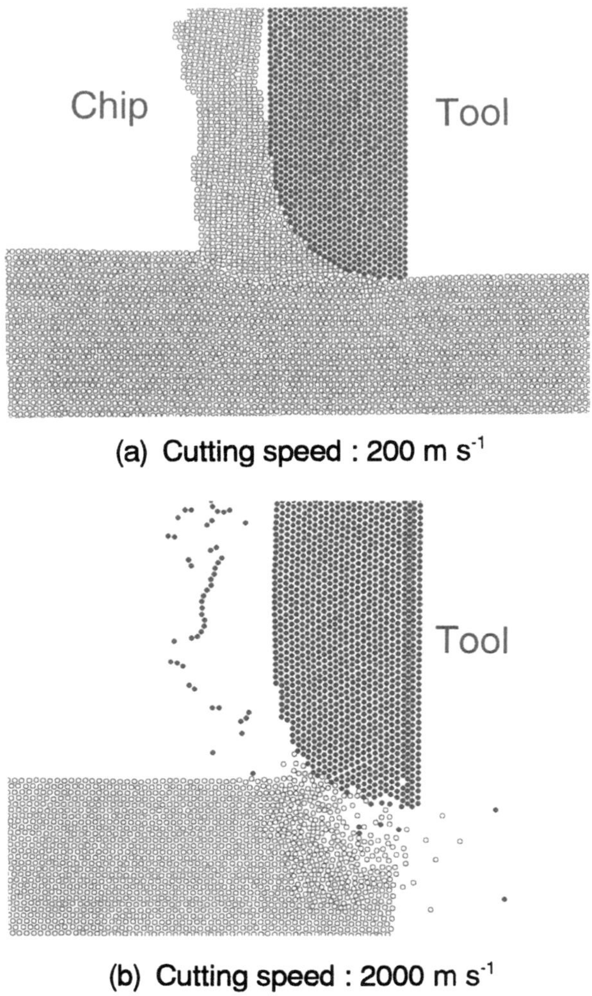

Fig. 12. Material removal under ultra high cutting speed (cutting edge radius : 5 nm, undeformed chip thickness : 1 nm).

References

1. N. Ikawa, R.R. Donaldson, R. Komanduri, W. Konig, P.A. McKeown, T. Moriwaki, I.F. Stowers: Ann. CIRP **40**, 587-594 (1991).
2. R.R. Donaldson, D.C. Tompson: Ann. CIRP **35**, 373-376 (1986).
3. Y. Kami, M. Yabuya, T. Shimizu: Proc. 4th Biennial Joint Warwick/Tokyo

Nanotechnology Symposium, Warwick, 1994.
4. H. Mizumoto, S. Arii, A. Yoshimoto, T. Shimizu, N. Ikawa: Ann. CIRP **44**, 523-526 (1996).
5. N. Ikawa, S. Shimada, H. Tanaka, G. Ohmori: Ann. CIRP **40**, 551-554 (1991).
6. J.F. Belak, I.F. Stowers: Proc. ASPE 1990 Annual Conference, Rochester, 1990, pp.259-262
7. T. Inamura, H. Suzuki, N. Takezawa: J. Jpn. Soc. Prec. Eng. **56**, 1480-1486 (1990) (in Japanese).
8. I.F. Stowers, J.F. Belak, D.A. Lucca, R. Komanduri, R.L. Rholar, T. Moriwaki, K. Okuda, N. Ikawa, S. Shimada, H. Tanaka: Proc. 1991 ASPE Annual Meeting, Santa Fe, 1991, pp.100-103.
9. S. Shimada, N. Ikawa, G. Ohmori, H. Tanaka, J. Uchikoshi: Ann. CIRP **41**, 117-120 (1992).
10. S. Shimada, N. Ikawa, H. Tanaka, G. Ohmori, J. Uchikoshi: Ann. CIRP **42**, 91-94 (1993).
11. S. Shimada, N. Ikawa, H. Tanaka, G. Ohmori, J. Uchikoshi: J. Jpn. Soc. Prec. Eng. **59**, 2015-2021 (1993) (in Japanese).
12. S. Shimada, N. Ikawa, H. Tanaka, J. Uchikoshi: Ann. CIRP **43**, 51-54 (1994).
13. S. Shimada, R. Inoue, J. Uchikoshi N. Ikawa: Proc. SPIE **2576**, 396-405 (1995).
14. S. Shimada, N. Ikawa, R. Inoue, J. Uchikoshi: Proc. 1995 ASPE Annual Meeting, Austin, 1995, pp.21-24.
15. W.G. Hoover, C.G. Hoover, J.F. Belak, I.F. Stowers, A.J. DeGroot: Thrust Area Report FY90, Lawrence Livermore National Laboratory, Livermore, 1990, pp.4-1-4-8.
16. N. Ikawa, S. Shimada: Proc. 3rd Int. Conf. on Production Eng. Kyoto, (1977) pp. 357-364.
17. N. Ikawa, S. Shimada, H. Tanaka: Nanotechnol. **3**, (1992).
18. S. Shimada, N. Ikawa, R. Inoue, J. Uchikoshi: Proc. JSPE 1996 Spring Meeting, Tokyo, 1996, pp.143-144 (in Japanese).

Molecular Dynamics Simulation of Tensile Deformation in Amorphous $Zr_{67}Ni_{33}$ Alloy

Tomoyasu Aihara, Jr. and Yoshiyuki Kawazoe

Institute for Materials Research, Tohoku University, 2-1-1 Katahira, Aoba-ku, Sendai 980-8577, Japan

Summary. Molecular dynamics simulation is performed to study the deformation and fracture processes in a Zr-Ni amorphous alloy. A Finnis-Sinclair type pair functional potential is used. Uni axial strain is applied for a nano-size rod with constant strain rate at 77 K. Tensile processes are traced by checking the stress-strain curve, energy change in potential and by examining snapshots of atom movement The discrete atomic model reproduces the elastic continuum behavior. The plastic deformation occurs by the viscous flow mechanism, and no work hardening is observed. The distribution of atomic level stress is analyzed.

Key Words. Molecular dynamics, Zr, Ni, Amorphous alloy, Plastic deformation

1. Introduction

Plastic deformation in amorphous alloys at temperatures well below the glass transition point was studied in detail for the first time by Masumoto and Maddin [1]. Experimentally, following features are reported for the deformation and fracture in the amorphous alloy. (1) It is a high strength alloy. (2) It exhibits no substantial work hardening. (3) The elastic modulus of an amorphous alloy is lower than that of a crystalline alloy. (4) The plastic deformation is confined to very thin shear bands along the direction of maximum shear stress. (5) Scanning electron microscopy (SEM) micrographs of a ductile fracture surface show a *vein* pattern. These mechanical behaviors of the amorphous alloys differ from those of crystalline alloys.

Dislocation-assisted and viscous flow models have been proposed to explain the macro and structural level behaviors. Gilman [2] has postulated the existence of crystal-like dislocations in amorphous structures by analogy between the inhomogeneous slip in amorphous alloys and that in crystalline alloys. He explained the deformation of amorphous alloys in terms of creation and annihilation of dislocations. Li [3] proposed a dislocation lattice which consists of regularly arranged edge dislocations. The shear modulus, the critical shear stress and localized inhomogeneous deformation were explained by cooperative dislocation motion.

On the other hand, in experiments, the viscous nature of the deformation in stress concentrated regions has been pointed out by Masumoto and Maddin [1]. The free-volume model of Spaepen [4] and the localized-shear transformation mechanism of Argon [5] are based on the concept of the atomic flow in viscous matter. Although these four models have succeeded in explaining a part of the problem, they are not well founded on experimental evidence.

Computer simulation is an effective tool to observe the atomic level dynamics in contrast to the difficulties of in situ high resolution microscopy observation. The atomic level simulations for single component amorphous metal deformation have been reported for two [6] and three [7, 8] dimensional Johnson potential systems by using the static relaxation method. Molecular dynamics (MD) simulations with Lennard-Jones [9] and Johnson potential [10] have been performed to study the anisotropy of the static structure of the amorphous metal after deformation.

To examine the atom level dynamics at finite temperature, time evolution of the system should be solved throughout the deformation procedure. However, previous simulations are not sufficient at this point. In this paper, we report the deformation process of a binary Zr-Ni amorphous alloy simulated by molecular dynamics. The stress-strain curve, atom level cross-sections and trajectories are analyzed to clarify the deformation process. Furthermore, the atom level energy distribution is calculated.

2. Simulation Method

The simulated $Zr_{67}Ni_{33}$ alloy system consists of 5760 Zr atoms and 2880 Ni atoms. MD calculation is begun by constructing C16 tetragonal crystal lattice cells ($a \times c \times a$: 6×20×6 lattice cells) with periodic boundary conditions for x, y and z directions. The center of mass of the system is located at the origin of the coordinate axes. The system is first heated well above the melting temperature (T_m = 1393 K), i.e., up to 2000 K, to obtain a homogeneous equilibrium liquid state. The system is quenched quasi-continuously to various object temperatures with a quench rate of 5×10^{13} K s^{-1}. A constant thermal expansion coefficient (2.659×10^{-5} K^{-1}) is assumed in this thermal history. The density of the system is set as 7.06 Mg m^{-3} at 300 K to agree with the experimental value [11]. After isothermal annealing at an object temperature for 10 ps, free surface boundary conditions are introduced for x and z directions. The configuration of surface atoms is relaxed during 10 ps at the same temperature. The system has a rod shape with the major axis parallel to the y direction. Uni axial strain is loaded up to 140% for this system. The strain is defined as the ratio of change in length ΔL to the original length L_0 = 10.56 nm. The stepwise 0.2% strain is applied per 0.2 ps by scaling the MD cell length in y direction (strain rate 1×10^9 s^{-1}). We use a Nosé constant-temperature (NVT) ensemble [12]. The equations of motion were solved using the fifth-order predictor corrector algorithm of Gear with a time step of 1 fs.

We use the pair functional potential suggested by Massobrio, Pontikis and Martin (MPM) [13]. The potential energy of the system is written as

$$E_{pot} = \sum_{\alpha=1}^{m} \sum_{i_\alpha}^{N_\alpha} \left[\sum_{\beta=1}^{m} \sum_{j_\beta (\neq i_\alpha)}^{N_\beta} A_{\alpha\beta} \exp\left[-p_{\alpha\beta} \left(\frac{r_{i_\alpha j_\beta}}{d_{\alpha\beta}} - 1 \right) \right] \right.$$

$$\left. - \sqrt{ \sum_{\beta=1}^{m} \sum_{j_\beta (\neq i_\alpha)}^{N_\beta} \xi_{\alpha\beta}^2 \exp\left[-2\, q_{\alpha\beta} \left(\frac{r_{i_\alpha j_\beta}}{d_{\alpha\beta}} - 1 \right) \right] } \right], \tag{2.1}$$

where $r_{i_\alpha j_\beta}$ is the distance between i_α and j_β atoms. The latter functional term represents many body interactions in metallic bonding matter. They fitted the potential to reproduce the thermodynamical and elastic properties of the crystal structures. In previous papers, we have reported the MD simulation of the $Zr_{67}Ni_{33}$ amorphous alloy with no external constraint by using this MPM potential. The simulated structure factor and the vibrational energy distribution in the amorphous alloy reproduce the experimental values well [14-20]. We found the simulated glass transition temperature of the $Zr_{67}Ni_{33}$ alloy to be 640 K, which is close to the experimental value of 652 K [14].

The macroscopic physical property of the matter is the average of the atom level quantities. For example, the macroscopic stress tensor of the system is defined as follows,

$$\sigma^{\alpha\beta} = \frac{1}{V} \sum_{i=1}^{N} \left[m_i v_i^\alpha v_i^\beta + \sum_{\substack{j=1 \\ j \neq i}}^{N} \frac{1}{2} F_{ij}^\alpha r_{ij}^\beta \right] \tag{2.2}$$

where α and β denote the components of the vector in Cartesian coordinates. m_i and $\boldsymbol{v}_i$ are the mass and velocity of atom i. V is the volume of the system. $\boldsymbol{F}_{ij}$ is the force on atom i due to atom j, which is calculated as

$$\boldsymbol{F}_{ij} = -\frac{dE_{pot}}{dr_{ij}} \frac{\boldsymbol{r}_{ij}}{r_{ij}}. \tag{2.3}$$

A negative value of the diagonal part of the tensor denotes normal tensile stress. Off-diagonal parts represent shear stress.

Some atom level physical quantities change with inhomogenity of structure in the matter [21]. The stress of the system is distributed to individual atoms. We defined the atom level stress energy as follows;

$$E_i^{\alpha\beta} = m_i v_i^\alpha v_i^\beta + \sum_{\substack{j=1 \\ j \neq i}}^{N} \frac{1}{2} F_{ij}^\alpha r_{ij}^\beta. \tag{2.4}$$

Since an atomic volume depends on local atom configuration in an amorphous structure, the dimension of the atom level quantity is energy per atom. Atom level potential energy for an atom is likewise defined as the summed terms in the largest brackets in (2.1).

3. Results

During the simulation of deformation we analyze the stress-strain curve. Each component of the true stress tensor is shown in Fig. 3.1 as a function of tensile strain. The tensile stress is plotted as positive values according to the engineering custom in contrast to the definition of (2.2). The *yy* component is the normal stress parallel to the tensile direction. First, the stress drastically increases up to 5% strain. Secondly, a stress plateau is observed from 5% to 95% strain. Thirdly, the stress decreases with strain over 95%. The first stage represents the elastic deformation. The simulated Young's modulus of the amorphous alloy under 3% strain is 43 GPa, which is lower than that of polycrystalline Zr (98 GPa) and Ni (200 GPa). The random structure in the amorphous alloy is the origin of the softening of the Young's modulus. The second stage indicates the homogeneous plastic deformation. The stress-strain curve indicates no work hardening. The average of plastic flow stress between 10 and 90% is 2448 MPa. The transition from elastic to plastic deformation is apparent in the present simulation. The third stage is the inhomogeneous plastic deformation. In the tensile experiment, the nominal stress is measured conventionally, which is defined as the ratio of tensile normal force (F) to the initial cross-section of the specimen perpendicular to the tensile direction (A_0). It is calculated as a product of true stress and factor $V / [(\Delta L + L_0) A_0]$. The simulated nominal stress-strain curve is similar to the experimental results [22].

As can be seen in Fig. 3.1, the shear stresses (*yz* and *zx*) and the normal stress perpendicular to the tensile direction (*zz*) do not depend on the strain. The fluctuations of these stresses are very small in comparison with normal stress in the tensile direction. The average of the shear stresses is nearly zero and the average of the *zz* stress takes a small compressive value of -34 MPa. The former means that the present simulation is performed with a macroscopic shear stress free condition. The latter is caused by the surface tension in the amorphous rod.

The potential energy change is shown in Fig. 3.2. At the first stage, the energy drastically increases by elastic deformation. At the second stage, the potential energy is proportional to the strain. The increment depends mainly on the free surface formation during homogeneous plastic deformation. It is small under the

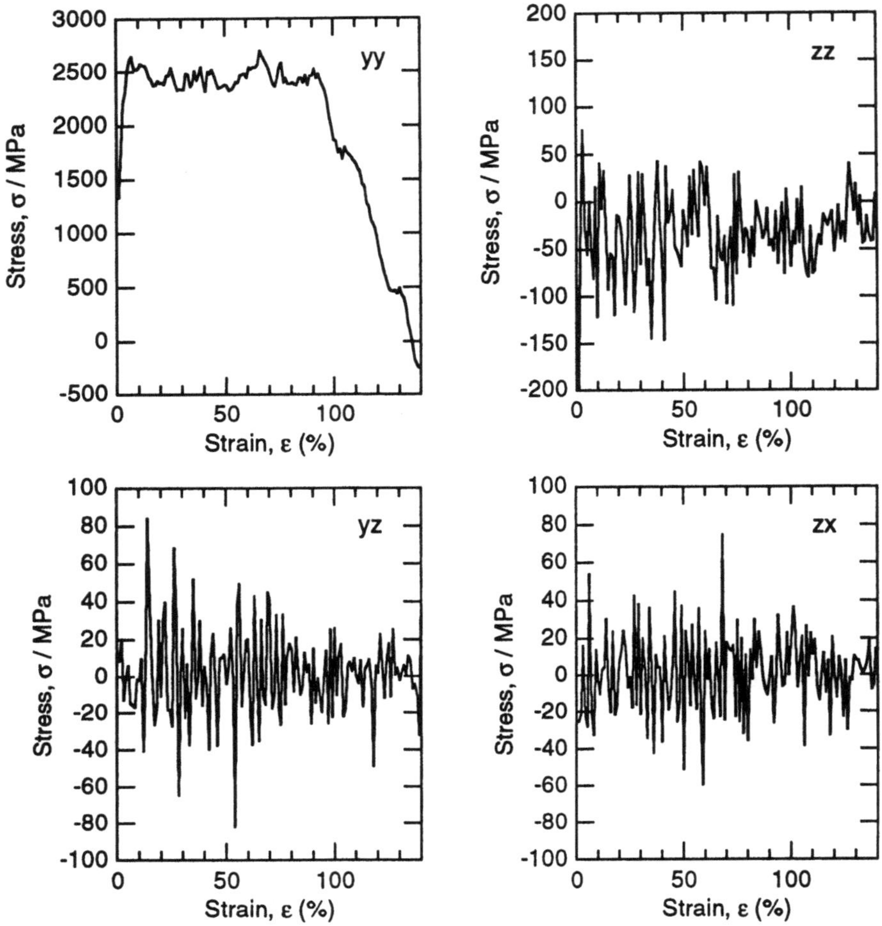

Fig. 3.1. True stress-strain curve estimated for $Zr_{67}Ni_{33}$ alloy. Tensile direction is parallel to the *y* axis. The *yy* and *zz* components are normal, and *yz* and *zx* are shear stresses, respectively.

inhomogeneous plastic deformation stage. The total potential energy change in the system is of the order of 1%.

Macroscopic shape change is defined as the sum of the atom configuration change. Fig. 3.3 shows snapshots of the *xy*-section at $z = 0$ during deformation. Ni atoms are indicated as dark gray spheres with 0.124 nm radius and Zr atoms as white ones with 0.160 nm radii. These are Goldschmidt radii. Very large elongation and completely ductile fracture are observed. The *xz* cross-section changes from a square to a circular shape with increase of the strain. The rod uniformly deforms below 40% strain, i.e. in the elastic stage and in the early plastic stage.

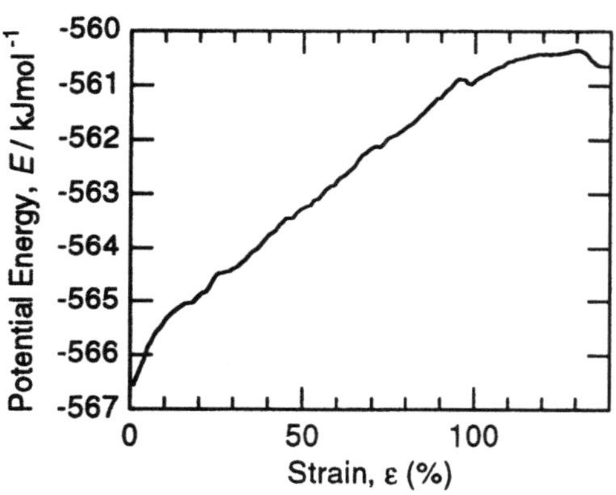

Fig. 3.2. Potential energy change in $Zr_{67}Ni_{33}$ alloy as a function of tensile strain.

Two neckings are observed beyond 40% strain. The neckings slightly develop in the homogeneous plastic deformation range. The right hand side necking is drastically drawn down to a point over 90% strain. This is the inhomogeneous plastic deformation. After the fracture at 132% strain, the ruptured spindle shape is quickly relaxed to a hemispherical shape by surface atom rearrangement. The position of the necking and the fracture depend on the initial condition of the random numbers. All the atoms are treated equally in the simulation, but they can not be predicted at first. Young's modulus and the flow stress do not depend on the initial condition. Apparently, we do not observe the vacancies in these snapshots. The gray and white patterns of the atom configurations do not change much during deformation. The system is not in a liquid state.

We analyze the deviation for each atom to clarify the atomistic deformation processes in detail. The deviations of atoms between $z = \pm 0.4$ nm are illustrated as a projection to the *xy*-plane in Fig. 3.4. The head of the needle indicates the initial position and the tip the final one. From 1 to 5% strain, the rod system elastically deforms in a continuum way. The atom deviation is symmetric as indicated by the gray arrows. The deviation is proportional to the *y* position of the atom. At the early stage of homogeneous plastic deformation, the atom deviation is basically proportional to the *y* position as indicated in 28- 32% snapshot. In the homogeneous deformation range, all the atoms contribute to the deformation. However, the atoms in the shaded area do not move in the *xy*-plane. The homogeneous plastic deformation occurs in a non-continuum way, and consists of cooperative motion of atoms on the order of the third neighbor atom distance. At the late stage of the homogeneous plastic deformation, the non-moving area increases as shown in the result of the 58-62% strain. The system is slightly necked, but all atoms contribute to the plastic deformation. Semi-macro scale shear motion is observed as indicated by the gray arrows. In the case of inhomogeneous plastic deformation, only the atoms at the necking area contribute to the plastic deformation, as shown in the result of the 108-112% strain. In case of the inhomogeneous deformation range, deformation is local at the atom level.

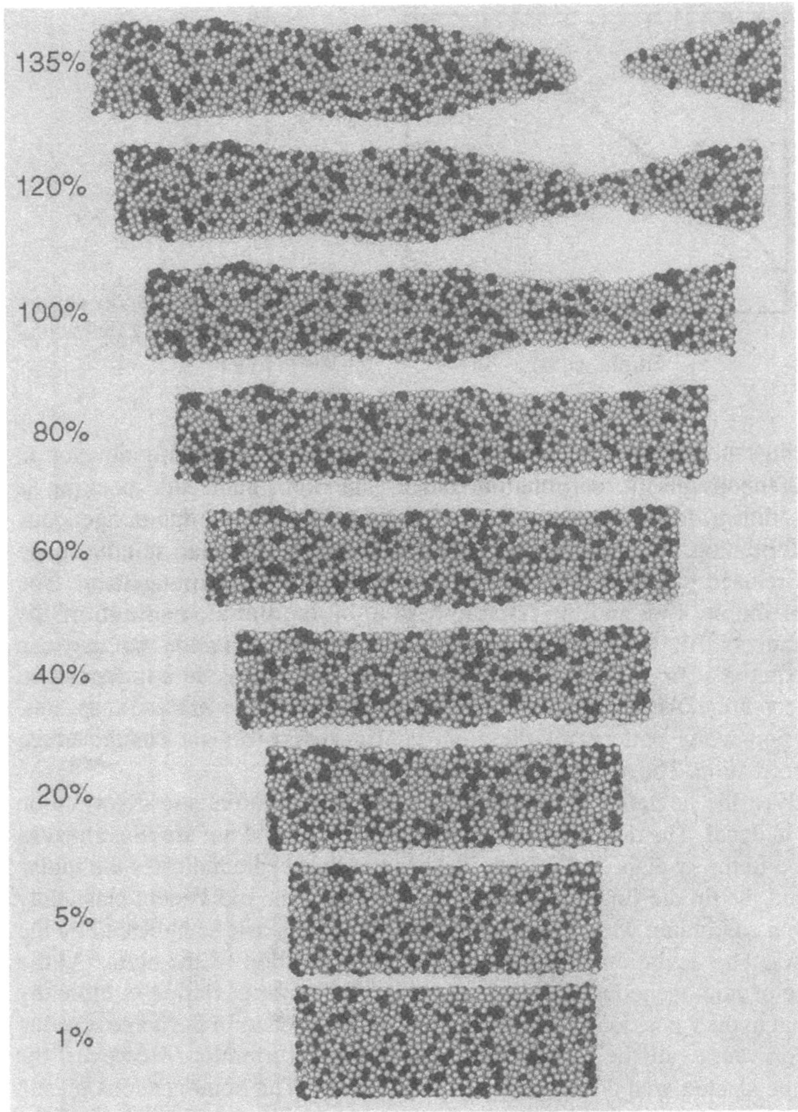

Fig. 3.3. Atom configuration change during deformation as snapshots of the xy-section. The x axis is vertical and the y horizontal in the page. The z axis is perpendicular to the page.

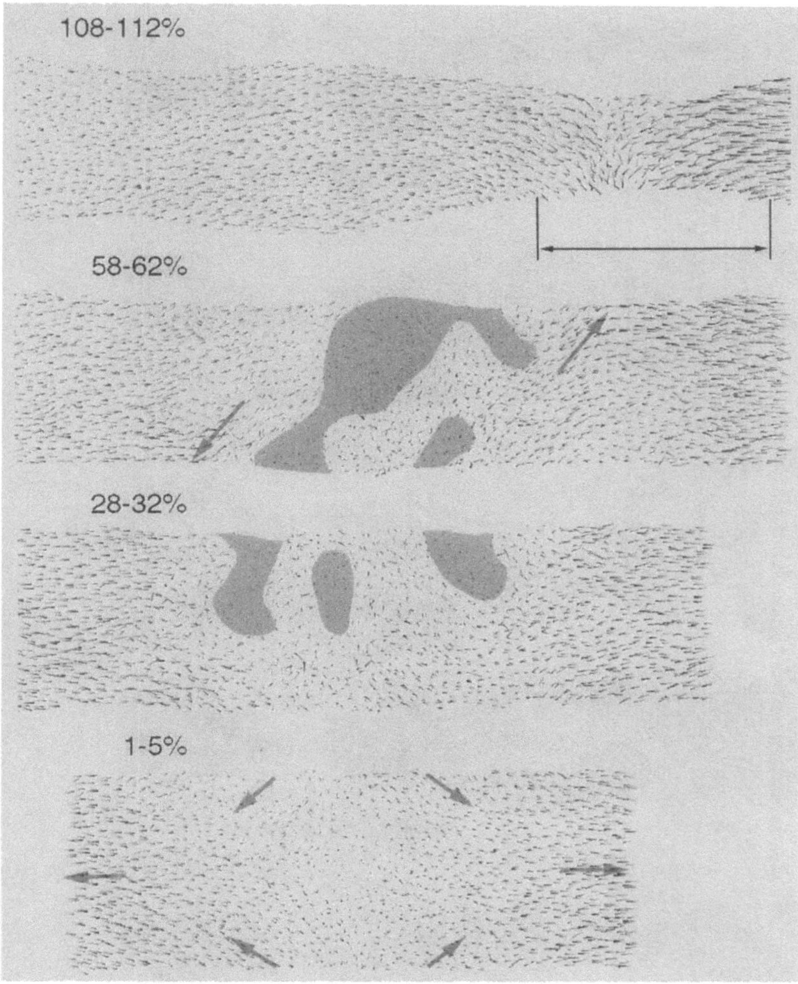

Fig. 3.4. The deviations of atoms during 4% stretching as a projection to the xy-plane. Shaded areas indicate less motion.

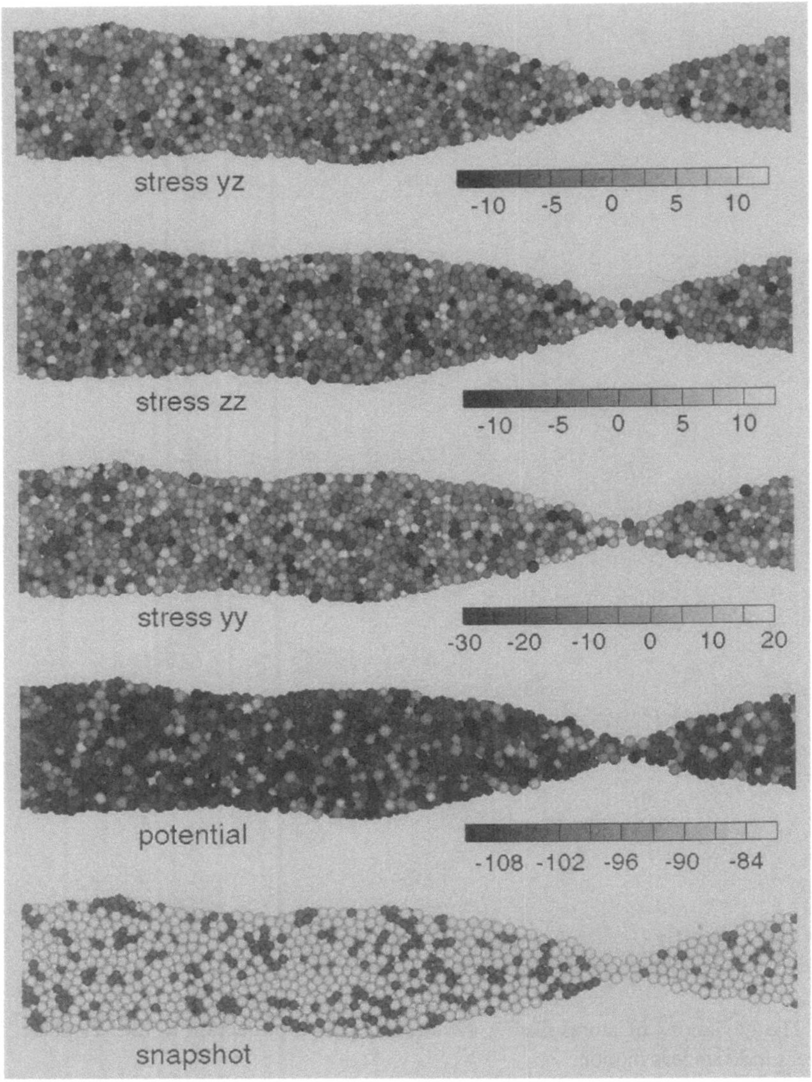

Fig. 3.5. Snapshot and configuration of atom level potential and stress energies at 130% strain. The *yy* and *zz* components are normal, and *yz* is shear stresses. The unit in the legends is 10^{-20} J/atom. A positive normal strain energy denotes compression.

The existence of inhomogeneity in the amorphous structure is visualized by mapping the atom level quantities. The distributions of atom level potential and stress energies at 130% strain are shown in Fig. 3.5 similar to the cross-section of Fig. 3.3. The positive normal strain energy denotes compression in this figure. Atom level fluctuation is very large in comparison with the averaged value. The standard deviation in atom level potential energy is 15 % of its average. On the other hand, the total potential energy change during 140 % strain is in the order of 1 %. There is no correlation between the atom level energy and local composition. At the high atom level potential energy region, the *yy* normal atom level stress energy denotes the compressive value. The atom level potential energy does not correlate to the atom level *yz* shear and *zz* normal stress energies at any stage of the deformation.

4. Discussion

The shape change in behavior of the simulated amorphous rod is similar to the ductile behavior in an Instron-type tensile test. However, the present simulation system does not directly correspond to the macroscopic mechanical tests, since the system size is quite different between experiment and simulation. The simulation represents the area with highly concentrated stress, in which the vein pattern is created. Because of limited computer power, the macroscopic inhomogenity can not be reconstructed by atom level simulation.

In general, the macroscopic strain and all of the atom level strains are equivalent for elastic deformation, but they are not equivalent for plastic deformation. In the case of a crystal, macroscopic plasticity occurs by propagation of dislocations in a periodic atom array. Two features should be considered for the amorphous alloy deformation. First, the amorphous structure is a random atom configuration. Secondly, the deformation is in a dynamic, non-steady and non-equilibrium state. We discuss the deformation in these terms.

The amorphous structure is uniform and isotropic. Accordingly, as can be seen in Fig. 3.3, no vacancies are apparent in the present simulation both in the as prepared state and the deformed state. We also made a computer generated video of the deformation process to clarify details of the atom dynamics. We could not identify the long-lived extended line defects such as dislocations in the video. The plastic deformation of the amorphous alloy does not originate from inhomogeneous defects such as dislocations or vacancy\ies. In the video, the major motion is thermal vibration and the minor motion is jumps with half order of the interatomic distance movement. The jump motion in the amorphous alloy also occurs under a static state [18]. However, the jump motion is more frequent in the strained condition. The jump motion changes the local atom configuration and causes the macroscopic plastic deformation.

Many local atom configurations exist in the amorphous state, in contrast to the crystalline state. Therefore the energy distribution of the amorphous alloy is wider than that of the crystal [20]. Slight structural modification is available in the random structure. The amorphous phase is a non-equilibrium state. In the static state, structure relaxes on very long time scales. The deformation process is also far from an equilibrium state. The stress accelerates the relaxation. It increases the local change of the atom configuration by jump motion and collective atom motion. Although the mobility is increased, it is not a liquid. No long range order hydrodynamic diffusion is observed. The yield behavior of the amorphous alloy is best described by a viscous flow and it differs from that of crystalline metals.

To clarify more exact information on deformation and fracture, it is desired to perform first principle molecular dynamics calculation on very large scales up to order of ns period by extremely powerful computers in the next decade.

5. Conclusions

A molecular dynamics simulation has been performed to study the deformation and fracture processes in Zr-Ni amorphous alloy. Very large elongation and completely ductile fracture are observed. There are no vacancies and no dislocations in the amorphous structure during the simulation. The deformation process consists of elastic, homogeneous plastic and inhomogeneous plastic stages. The stress-strain curve indicates no work hardening. The amorphous alloy elastically deforms in a continuum way. The Young's modulus of the amorphous alloy is lower than that of polycrystalline metals. The neckings develop slightly in the homogeneous and drastically in the inhomogeneous plastic deformation range. The homogeneous plastic deformation occurs in a non-continuum way and consists of cooperative motion of atoms on the order of the third neighbor atom distance. The stress increases the local change of the atom configuration by jump motion and collective atom motion. The yield behavior of the amorphous alloy is a kind of viscous flow and different from that of crystalline metals. There is no correlation between the atom level energy and local composition.

Acknowledgments

The authors wish to thank the Information Science Group of IMR, Tohoku University for support of the HITAC S-3800/380 system. This research was supported in part by a Grant-in-Aid for Scientific Research from the Ministry of education, Science and Culture of Japan on Priority Area of "Cooperative Phenomena in Complex Liquids".

References

1. T. Masumoto R. Maddin: Acta Metall. **19**, 725 (1971).
2. J. J. Gilman: J. Appl. Phys. **44**, 675 (1973).
3. J. C. M. Li: *Frontier in Materials Science*, (Marcel Dekker, New York, 1976) p.527.
4. F. Spaepen: Acta Metall. **25**, 407 (1977).
5. A. S. Argon: Acta Metall. **27**, 479 (1979).
6. K. Maeda, S. Takeuchi: Phys. Status Solidi A **49**, 685 (1978).
7. R. Yamamoto, H. Matsuoka, M. Doyama: Phys. Status Solidi A **51**, 163 (1979).
8. T. Egami, K. Maeda V. Vitek: Phil. Mag. A **41**, 883 (1980).
9. T. Tomida T. Egami: J. Appl. Phys. **69**, 5451 (1991).
10. T. Tomida T. Egami: Phys. Rev. B **48**, 3048 (1993).
11. Z. Altounian, J. O. Strom-Olsen: Phys. Rev. B **27**, 4149 (1983).
12. S. Nosé: J. Chem. Phys. **81**, 511 (1984).
13. C. Massobrio, V. Pontikis, G. Martin: Phys. Rev. B **41**, 10486 (1991).
14. T. Aihara, Jr., K. Aoki, T. Masumoto: Scr. Metall. **28**, 1003 (1993).
15. T. Aihara, Jr., K. Aoki, T. Masumoto: Mater. Sci. Eng. **A179-180**, 256 (1994).
16. T. Aihara, Jr., K. Aoki, T. Masumoto: Mater. Trans. JIM **36**, 399 (1995).
17. T. Aihara, Jr., T. Masumoto: J. Phys. Condens. Matter **7**, 1525 (1995).
18. T. Aihara, Jr., Y. Kawazoe, T. Masumoto: Mater. Trans. JIM **36**, 835 (1995).
19. T. Aihara, Jr., Y. Kawazoe, T. Masumoto: Sci. Rep. RITU A **41**, 1 (1995).
20. T. Aihara, Jr., Y. Kawazoe, T. Masumoto: Mater. Sci. Eng. B **37**, 201 (1996).
21. T. Aihara, Jr., Y. Kawazoe: Prog. Theor. Phys. Supp. **126**, 355 (1997).
22. Y. Kawamura, T. Shibata, A. Inoue, T. Masumoto: Appl. Phys. Lett. **69**, 1208 (1996)

Molecular Dynamics Study on Fracture Mechanism of Fe-Amorphous Metal (J Integral near Mode I Crack Tip)

Keiko Nakatani[1], Akihiro Nakatani[2], and Hiroshi Kitagawa[2]

[1]Department of Aerospace Engineering, College of Engineering, Osaka Prefecture University, 1-1, Gakuen-cho, Sakai, Osaka 599-8531, Japan,
[2]Department of Adaptive Machine Systems, Graduate School of Engineering, Osaka University, 2-1, Yamadaoka, Suita, Osaka 565-0871, Japan

Summary. When molecular dynamics (MD) simulation of crack propagation and fracture is carried out, it is very common that the boundary conditions are determined by the atomic displacement which correspond to a stress intensity factor K based on linear elasticity with the initial elastic properties of the material. However, in our previous studies, it was shown that a very high strain occurs near the crack tip [1], and the initial mechanical properties are changed as the elastic constants decrease because of non-elastic deformation under large stress concentration [2]. Therefore, the assumption of a linear elastic condition (small scale yielding) cannot be realized any longer and the boundary condition based on a linear elastic solution is no longer applicable. In this paper, the J integral, which is one of the representative mechanical parameters that often gives the criterion of crack propagation in continuum mechanics, is evaluated in an atomistic Fe-amorphous model in order to investigate quantitatively the mechanical state of the damaged mechanical field near the crack tip. J is actually evaluated by J^* which is advantazeous in MD simulations. The molecular dynamics simulations of mode I crack propagation in Fe-amorphous metal under the boundary condition which controlled displacements of boundary atoms are carried out for the purpose of evaluating the damaged mechanical field near the crack tip, and J^* is evaluated. As a result, it is shown that the change of mechanical properties is concerned with the material non-linearity and the blunting shape, and that it is predicted quantitatively by using the J* integral which is evaluated by choosing the appropriate integral region.

Key Words. Fracture, J integral, computational mechanics, molecular dynamics, Fe amorphous metal, mode I crack, damage

1. Introduction

To study the fracture mechanism of materials, it is important that the deformation, stress and strain near the crack tip are evaluated correctly.

Fracture mechanics begins with the theory based on energy balance proposed by Griffith [3]. Orowan [4] and Irwin [5] improved it and extended it to quasi-brittle fracture. Moreover, after Irwin introduced the concept of a stress intensity factor K and an energy release rate $\mathcal{G}$ [6] and the concept of small scale yielding was established, the fracture mechanics had been developed. Since then, the crack opening displacement (COD) [7] and J integral [8] which are relevant for large scale were proposed yielding, and fracture mechanics is now recognized as one of the indispensable tools in engineering.

In the case of small scale yielding, the relationship between K, which is a parameter for the distribution of stress and of strain near the crack tip, and the crack opening shape has been analyzed [9]. However, the properties of the field near the crack tip do not always satisfy the condition of small scale yielding necessarily. When an elasto-plastic region develops in such a situation, the parameter based on the linear fracture mechanics is no longer applicable. The parameter which is applied to such a state is significant. In such large scale yielding, the most representative parameters for the evaluation of crack growth property are the method which is based on the J integral [10, 11]. Moreover, some parameters which can apply to not only static but dynamic crack propagation are proposed [12] – [14]. Especially, Paris *et al.* [10], Hutchinson and Paris [11], Rice and Sorensen [15] and Shih *et al.* studied the applicability of the non-dimensional parameter T_J which is called the tearing modulus and represents the gradient of the J integral-resistance curve, in order to characterize theoretically the ductile crack propagation by using the J integral. Many parameters which describe the mechanical state near the crack tip have been proposed, as mentioned above.

On the other hand, when we study criteria of some essential phenomena of fracture by means of atomic level simulation, we need to know the values of these macroscopic parameters from the results of atomistic simulation.

For simplicity, external forces or displacement of the boundary atoms which are located near the surface in the atomic region are often controlled by using the linear elastic solution. However, our previous study [1] shows that very high strain occurs near the crack tip, and the initial mechanical properties are changed, i.e. the elastic constant decreases because of non-elastic deformation under a large stress concentration [2]. So, the assumption of a linear elastic condition (small scale yielding) is not applicable at a high loading level. In other words, the stress intensity factor K, which is given as the boundary condition based on a linear elastic solution, can no longer correctly express the phenomenon of the mechanical field near the crack tip.

In this paper, the J integral, which is one of representative mechanical parameters as the criterion of crack propagation in continuum mechanics, is evaluated in an atomistic simulation. In order to avoid the difficulty of evaluation, J is transformed to J^* which can be applied to the atomic model. The molecular dynamics simulations of mode I crack propagation in Fe-amorphous metal under the boundary condition which is given by controlled displacements of boundary atoms are carried out for the purpose of evaluating the mechanical field near the crack tip. Moreover, it is shown that the mechanical state surrounding the crack tip by the using the J* integral can be predicted quantitatively.

2. Analysis Model and Method

2.1 Molecular Dynamics

In a molecular dynamics simulation [16], the behavior of atoms which construct the simulation system is supposed to be governed by classical mechanics based on the Newtonian equation of motion as follows;

$$m^{(\alpha)}\frac{d^2\boldsymbol{q}^{(\alpha)}}{dt^2} = -\frac{\partial\Phi}{\partial\boldsymbol{q}^{(\alpha)}} \equiv \boldsymbol{F}^{(\alpha)}(t) \qquad (\alpha = 1, 2, \cdots, N), \tag{2.1}$$

where $m^{(\alpha)}$ is the mass of the α-th atom, $\boldsymbol{q}^{(\alpha)}, (\alpha = 1, 2, \cdots, N)$ is the position vector, and Φ is the potential energy of system. $\boldsymbol{F}^{(\alpha)}(t)$ represents the atomic interaction from other atoms at a time t and it is determined by the arrangement of atoms.

In the simulation, the initial value problems to solve the second order simultaneous differential equations (2.1) are calculated. Initial velocities are given to satisfy the Maxwellian distribution corresponding to the temperature of the system. In this study, the MD simulations are carried out under the (NVE) ensemble.

2.2 Crack Models and Method of Analysis

The Fe-amorphous metal structure generated by MD simulation of the process of heating and rapid-quenching mentioned in our previous paper [17] is regarded as a unit cell. The cylindrical region is cut from the atomic system which consists of the unit cells arranged cyclically. The model near the crack tip consists of 22 251 atoms. The radius and the thickness along the z direction are $r = 69.9$Å and $l_z = 17.6$ Å, respectively. The radius is four times as long as the side of the unit cell. The simulation of uniaxial tensile deformation [2] produces the elastic properties of the material; Young's modulus E is 146.4 GPa and the Poisson ratio ν is 0.26. The periodic boundary condition is applied along the z direction.

In this analysis, ten different models, in which a mode I crack is introduced by a displacement field based on linear elastic solutions corresponding to the stress intensity factor K_{I}, are named from model C to L, which are shown in Table 2.1.

Within the linear elasticity region, the value of the J integral is equivalent to the energy release rate $\mathcal{G}$. It is expressed by

$$J = \mathcal{G} = \frac{(1-\nu^2)K_{\mathrm{I}}^2}{E} \tag{2.2}$$

under the condition of plane strain. The J integrals which correspond with the stress intensity factors are also shown in Table 2.1. As mentioned in a later section, K_{I} and J in Table 2.1 are nominal values which differ from true values. The initial arrangement of model C is illustrated in Fig.2.1.

Table 2.1. Analytical models

Model	K_{I} (MPa$\sqrt{\mathrm{m}}$)	$J = \mathcal{G}$ (Pa·m)	Model	K_{I} (MPa$\sqrt{\mathrm{m}}$)	$J = \mathcal{G}$ (Pa·m)
C	0.6062	2.337	D	0.7072	3.181
E	0.7578	3.652	F	0.8588	4.691
G	1.0103	6.492	H	1.2629	10.144
I	1.5155	14.608	J	1.6670	17.675
K	1.7681	19.883	L	1.9196	23.437

Table 2.2. Condition of calculation

Number of atoms	22 251
Size of initial cell	$r = 69.9$ Å, $l_z = 17.6$ Å
Temperature	300K
Potential	Finnis-Sinclair potential
Time increment	2.0×10^{-15} s
External stress	1.0×10^{5} Pa

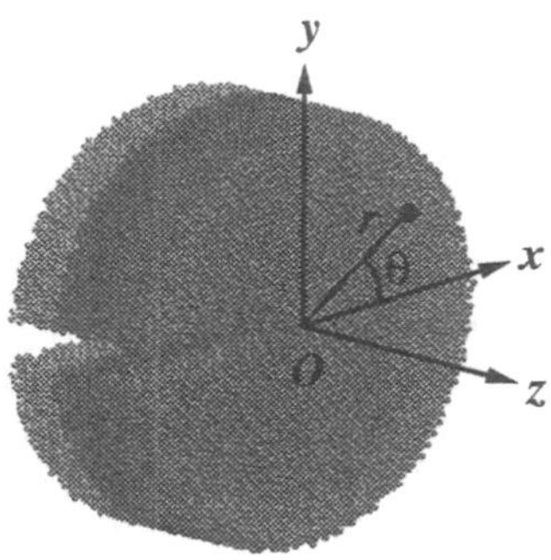

Fig. 2.1. Initial arrangement of model C.

MD simulation is carried out for relaxation during 8 ps under the condition in which 2206 atoms within the range $r_c = 3.57$ Å from the outside surface are fixed. Subsequent MD simulation is accomplished under the condition shown in Table 2.2, and potential energy density $W^{(\alpha)}$, stress $\sigma_{ij}^{(\alpha)}$, which is evaluated by the expression given in our previous paper [17], and displacement $u_i^{(\alpha)}$, at a location of each atom α are calculated by taking a time average during 1 ps.

The bulk model (model B), which has no crack, and the surface model (model S), in which there is no atomic interaction across the xz-plane, are calculated by the same procedure. They will be used as reference models.

The crack propagation analysis is also carried out using the same crack model. This simulation is termed the "loading model" for later reference. The relaxation calculation is performed at $K_{\mathrm{I}}^{init} = 0.758$ MPa$\sqrt{\mathrm{m}}$. After relaxation, the displacements of boundary atoms are changed as K_{I} gradually increases as $K_{\mathrm{I}} = K_{\mathrm{I}}^{init} + \dot{K}_{\mathrm{I}} \cdot t$ (,where $\dot{K}_{\mathrm{I}} = 0.505 \times 10^{10}$ MPa$\sqrt{\mathrm{m}}\cdot$ s^{-1}).

3. J Integral of Atomic Model

3.1 Definition of J^*

The J integral [18] is a representative parameter of the mechanical field near the crack tip, and it is defined in a continuum body as a line integral along a path surrounding the crack tip as follows,

$$J = \int_{\Gamma} \left(W\delta_{1j} - \sigma_{ij}\frac{\partial u_i}{\partial x_1} \right) n_j d\Gamma, \tag{3.1}$$

where W , σ_{ij}, and u_i are strain energy density, stress, and displacement, respectively. It is, however, difficult to estimate the line integral directly by using molecular dynamics simulations of the atomic model. By introducing a function q, whose values are 0 and 1 on the Γ_1 and Γ_2, respectively, into region S^*, which is surrounding by the closed path Γ^* as shown in Fig. 3.1, another form of integral J^* is considered as follows,

$$\begin{aligned} J^* &= \int_{\Gamma^*} \left(W\delta_{1j} - \sigma_{ij}\frac{\partial u_i}{\partial x_1} \right) q n_j \mathrm{d}\Gamma \\ &= \iint_{S^*} \left[\left(W\delta_{1j} - \sigma_{ij}\frac{\partial u_i}{\partial x_1} \right) q \right]_{,j} \mathrm{d}S. \end{aligned} \tag{3.2}$$

It is proven that J^* is equal to J by using the definition of W, the relationship of strain versus displacement, symmetry of stress, and balance equations. The mechanical state in the vicinity of the crack tip is studied here by direct calculation of,

$$J^* = \iint_{S^*} \left[\left(W\delta_{1j} - \sigma_{ij}\frac{\partial u_i}{\partial x_1} \right) \right] \frac{\partial q}{\partial x_j} dS. \tag{3.3}$$

It is difficult to evaluate displacement, strain, and stress, which are necessary for the estimation of J of amorphous metal, because they oscillate by the randomness of the atomic level. To avoid this difficulty, we suppose that the object values at point P are obtained by taking an average over atoms which

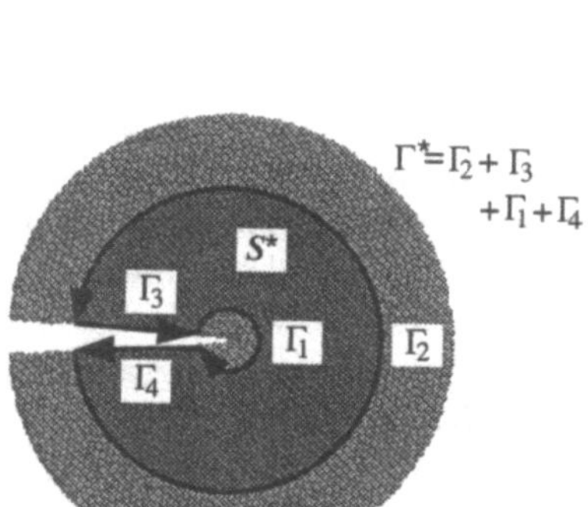

Fig. 3.1. Integral path for J* integral of atomic model of amorphous metal.

exist within a given region. The average is not taken by simple summation, but taken by weighted summation where a weighted function is used. Because the function must be decreasing with increasing distance from point P, we choose a normal distribution function which has appropriate variance.

3.2 Estimation of J^* in Atomic Model

By assumption that W is calculated by the change of potential energy density $W^{(\alpha)}$, J^* is calculated by,

$$J^* = \sum_{\alpha \mathrm{inS}^*} \left[\left(W^{\mathrm{C}(\alpha)} - W^{\mathrm{O}(\alpha)} \right) \delta_{1j} - \left(\sigma_{ij}^{\mathrm{C}(\alpha)} - \sigma_{ij}^{\mathrm{O}(\alpha)} \right) \frac{\partial u_i}{\partial x_1} \right] \frac{\partial q}{\partial x_j} V^{(\alpha)} \frac{1}{l_z}, \tag{3.4}$$

where $V^{(\alpha)}$ is the volume occupied by the α-th atom, the values with superscript B, C, and S indicate the results of models B, C, and S, respectively. The values with superscript O are the reference states which are introduced in order to remove the influence of initial stress and the increasing free surface energy by introducing a crack and to evaluate the increasing pure stress and pure strain energy as follows,

$$\sigma_{ij}^{O(\alpha)} = \sigma_{ij}^{B(\alpha)} \tag{3.5}$$

$$W^{O(\alpha)} = \begin{cases} W^{S(\alpha)} & (x < 0) \\ W^{B(\alpha)} & (x \geq 0). \end{cases} \tag{3.6}$$

The weighted function w is adopted as the 3-dimensional normal distribution whose origin is set at the evaluation point and the standard deviation is determined as the length of the cut-off distance. Displacement gradients are calculated by the concept of distribution as follows,

$$\frac{\partial u_i}{\partial x_1}(w) = -\int_{-\infty}^{\infty}\int_{-\infty}^{\infty}\int_{-\infty}^{\infty} u_i \frac{\partial w}{\partial x} \mathrm{d}x\mathrm{d}y\mathrm{d}z. \tag{3.7}$$

The following equation is actually used instead of (3.7) .

$$\frac{\partial u_i}{\partial x_1} = -\sum_{\alpha=1}^{N} u_i^{(\alpha)} \frac{\partial w}{\partial x} V^{(\alpha)} \tag{3.8}$$

4. Results

4.1 The Potential Energy, Stress and Displacement

The potential energy density $W^{C(\alpha)} - W^{O(\alpha)}$, the stress $\sigma_y^{C(\alpha)} - \sigma_y^{O(\alpha)}$, the displacement $u_x^{(\alpha)}$, and the displacement gradient $du_x^{(\alpha)}/dx$ which are evaluated by MD simulation for atoms existing near the x axis (-3 Å$< y < 3$ Å) and the linear elastic solutions for the position of the same atoms are shown in Fig. 4.1. The tendencies of the distributions obtained by these MD simulations agree very well with the linear elastic solution except in the vicinity of the crack tip.

4.2 Change of Stress Distribution by Loading

The distributions of hoop stress σ_θ on $y = \tan\theta x$ obtained in the "loading model" are shown in Fig. 4.2. When K_I is small, radial stress almost agrees with the linear elastic solution except near the crack tip. On the other hand, as K_I increases, the stress near the crack tip decreases. The region in which it does not follow the linear elastic solution expands. However, the distribution of stress, the except near the crack tip, agrees with the linear elastic solution. The treatment as an elastic body is valid at a few angstroms from the crack tip even in the case of $K_\mathrm{I} = 1.516$ MPa$\sqrt{\mathrm{m}}$.

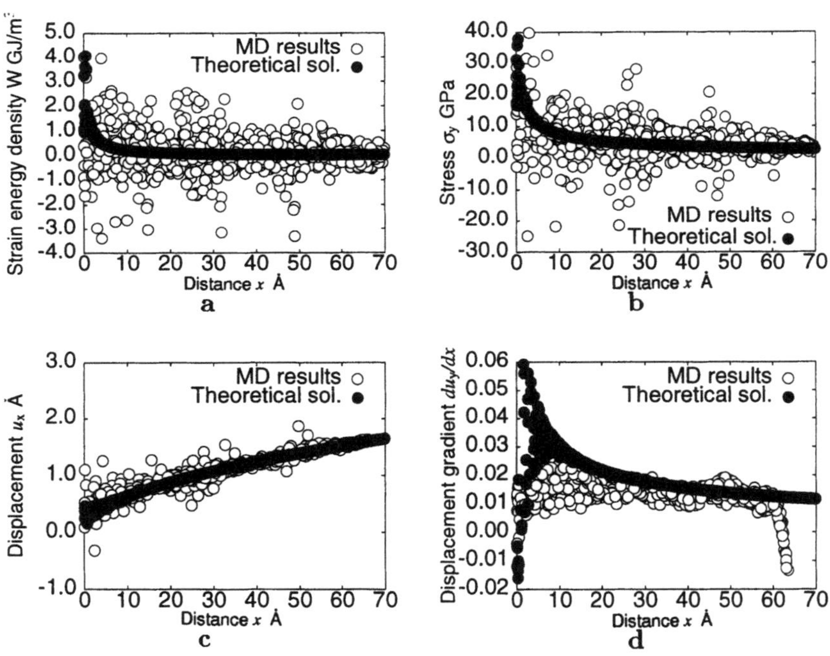

Fig. 4.1. Mechanical values by using results of MD analysis (Model C). **(a)** Potential energy density $W^{C(\alpha)} - W^{O(\alpha)}$, **(b)** Stress $\sigma_y^{C(\alpha)} - \sigma_y^{O(\alpha)}$, **(c)** Displacement $u_x^{(\alpha)}$, **(d)** Displacement gradient $\mathrm{d}u_x^{(\alpha)}/\mathrm{d}x$.

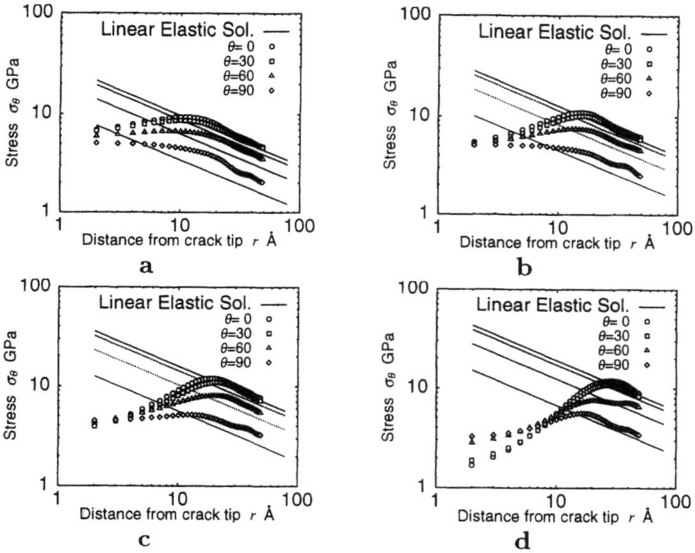

Fig. 4.2. Radial stress near the crack tip. **(a)** $K_{\rm I} = 0.768$ MPa$\sqrt{\rm m}$, **(b)** $K_{\rm I} = 1.010$ MPa$\sqrt{\rm m}$, **(c)** $K_{\rm I} = 1.263$ MPa$\sqrt{\rm m}$, **(d)** $K_{\rm I} = 1.516$ MPa$\sqrt{\rm m}$.

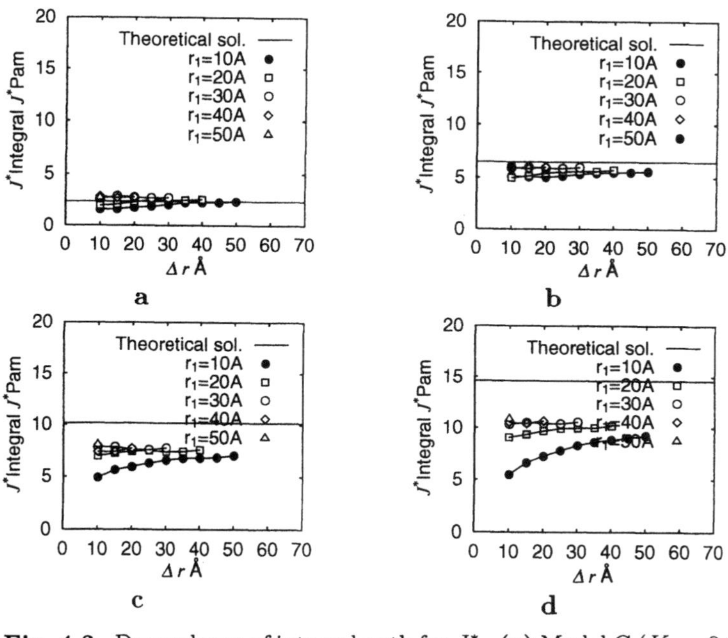

Fig. 4.3. Dependence of integral path for J^*. **(a)** Model C ($K_{\rm I} = 0.6062$ MPa$\sqrt{\rm m}$), **(b)** Model G ($K_{\rm I} = 1.0103$ MPa$\sqrt{\rm m}$), **(c)** Model H ($K_{\rm I} = 1.2629$ MPa$\sqrt{\rm m}$), **(d)** Model I ($K_{\rm I} = 1.5155$ MPa$\sqrt{\rm m}$).

4.3 J* Integral under Large Mode I Deformation

The dependencies on the J* integral path for the 4 models (C, G, H, and I), which are subject to different levels under mode I deformation are shown in Fig. 4.3. The J^* values based on the linear elastic solution evaluated by (2.2) are shown as solid lines in the figure. In the cases where K_I values are small (Fig. 4.3a,b), the dependencies on the J* integral path are almost negligible. As K_I becomes large (Fig. 4.3c,d), and r_1 is small, the J^* value decreases remarkably.

By comparing these results and Fig. 4.2, discussed in subsection 4.2, a remarkable decrease of the J* integral value under the condition where r_1 is small occurs when the J* integral path goes through the nonlinear region. However, even in the case where the region near the crack tip becomes a nonlinear state, if only the J* integral path is taken far from the crack tip (within the elastic region), it is basically independent of the J* integral path.

The J* integral value is better converged when a larger width of integral path, which is smaller than the value predicted by (2.2) under large deformation, is chosen.

This means that the nominal value is no longer as large as the true value, because of nonlinearity caused by deformation and damage near the crack tip.

4.4 Load Level Dependency of Strain Energy

The relation between the stress intensity factor K_I and the strain energy for all 10 crack models analyzed which receive different loads, are shown in Fig. 4.4. The solid curve drawn in the figure is the fitting curve formulated as $E_s = A \cdot K_\mathrm{I}^2$, and it represents the values predicted from the linear elastic solution. A constant A is selected so that the solid curve goes through the origin and the point for model C. Here, the point for model C is taken to be the mechanical values near the crack tip based on the linear elastic solution. According to Fig. 4.4, when K_I is small, the strain energies obtained as the result of MD simulation agree with the values predicted by the linear elastic solution.

However, when K_I is larger than the value for model G, they are smaller than the value predicted by the linear elastic solution. In the case of model H, the J* integral, which is calculated by taking $r_1 = 10$ Å, i.e. the integral path near the crack tip, begins to deviate from the results of other paths as shown in Fig. 4.3c. In model I, this tendency is more remarkable. Although mechanical properties are changed in the region near the crack tip where material is damaged by high stress, we can determine the mechanical state surrounding the crack tip by using the J* integral, which is calculated by choosing the appropriate integral path to avoid the damage zone.

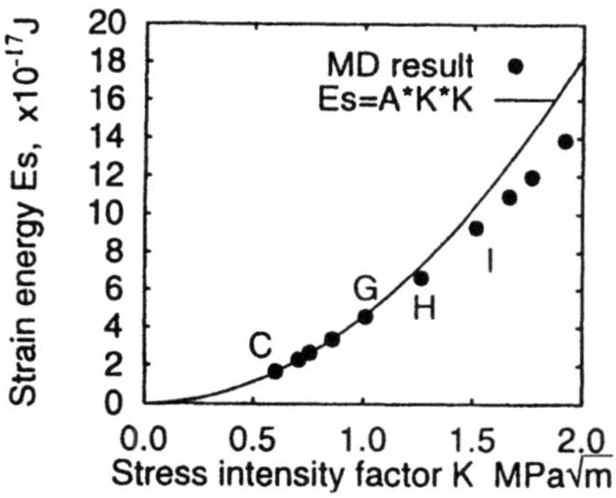

Fig. 4.4. Loading level dependence of strain energy.

5. Conclusions

Atomic Fe-amorphous models under mode I loading were analyzed by means of molecular dynamics (MD) simulation, and the J integral values which are representative parameters for mechanical properties around a crack tip were calculated, in order to evaluate the true mechanical fields near the crack tip.

Near the crack tip field, the mechanical properties were changed by non-linearity of material properties and atomic rearrangement of cleavage and blunting.

We were able to show that the J integrals were evaluated as J^* and that the mechanical state which is damaged by the change of internal mechanical properties was evaluated in detail by J^*, which is calculated on an appropriate integral region which avoids the process zone.

References

1 . K. Nakatani, H. Kitagawa, A. Nakatani: Prep. Jpn. Soc. Mech. Eng., **97**-1, 164 (1997).
2 . K. Nakatani, H. Kitagawa, A. Nakatani: J. Soc. Mat. Sci. Jpn., **46**-3, 244. (1997).
3 . A. A. Griffith: Philo. Trans. Roy. Soc., **221**, 163 (1920).
4 . E. Orowan: Rep. Progr. Phys., **12**, 185(1949).
5 . G. R. Irwin: Fract. Met., 152 (1948) ASM.
6 . G. R. Irwin: J. Appl. Mech., **24**-3, 361 (1957).
7 . A. A. Wells: British Welding Res. Ass. Rept., M13/63 (1963).
8 . J. R. Rice: (ed.by H. Liebowitz,) *Fracture II*, (Academic Press,1968) 213.
9 . J. R. Rice, M. A. Johnson: *Inelastic Behavior of Solids*, (McGraw-Hill, 1970) 641.

10 . P. C. Paris, *et al.*: ASTM STP, 668, 5 (1979).
11 . J. W. Hutchinson, P. C. Paris: ASTM STP, 668, 37 (1979).
12 . M. Saka: *et al.*, Trans. Jpn. Soc. Mech. Eng., **47**, A, 1301 (1981).
13 . M. Watanabe: Trans. Jpn. Soc. Mech. Eng., **47**, A, 406 (1981).
14 . M. Watanabe: Trans. Jpn. Soc. Mech. Eng., **48**, A, 1226 (1982).
15 . J. R. Rice, E. P. Sorensen: J. Mech. Phys. Solids, **26**, 163 (1978).
16 . S. Nośe: Prog. Theor. Phys. Suppl., **103**, 1 (1991).
17 . K. Nakatani, H. Kitagawa, A. Nakatani: Trans. Jpn. Soc. Mech. Eng., **62**-595, 847 (1996).
18 . J. R. Rice: Trans. ASME, J. Appl. Mech., 287 (1967).

Part 3

Criteria for Fracture

Instability Analyses of Stress-Induced Phase Transformation

Yoji Shibutani

Department of Mechanical Eng., Osaka University, Suita 565 Japan

Summary. The onset of stress-induced phase transformations has so far been predicted macroscopically by bifurcation conditions described in both thermodynamics and continuum mechanics. The soft mode phonon has been related to it from the microscopic lattice instability also. In the present paper, the scale-dependent relation between the two instability analyses is studied by both the extended macroscopic bifurcation conditions with atomistically estimated elastic constants and phonon calculations by molecular dynamics simulations. Moreover, fluctuations could play a main role in driving the transition and result in dissipation of energy. Therefore, the dynamics of the collective system of atoms during transformation are observed to determine the specific spatio-temporal pattern formation under the basic metallurgical mechanism of a slip system.

Key Words. Stress-induced phase transformation, molecular dynamics, bifurcation, soft mode phonon, fluctuation, pattern formation

1. Introduction

The onset of stress-induced phase transformations has so far been predicted as breaking bifurcation conditions derived from the positive definiteness of free energy in the whole system from phenomenological thermodynamics [1] and continuum mechanics [2]. Under a specific condition of an isothermal process, the free energy is equivalent to the strain energy stored in a deformed body [2]. Thus, breaking the positive definiteness of the strain energy also causes instability such as the plastic instability. Yip and co-workers [3 – 7] have studied this by introducing the extended bifurcation conditions involving the applied external stress [4] with the aid of the molecular dynamics simulations under the Parrinello and Rahman [8, 9] constant applied stress ensemble. Their results, from a single bcc crystal under a uniaxial tension, show the basic well known transition mechanism (Burgers' relation [10]) from a bcc to a hcp lattice structure, which involves a combination of atom shuffling and homogeneous shear [3]. Later, their proposed criteria under the finite deformation were confirmed [7].

On the other hand, breaking the positive definiteness of force constants defined as the second derivative of the employed interatomic potential according to atomic displacement has been discussed as the lattice instability [11] and it leads to a soft mode phonon [12]. Although the inevitable scale gap between the former bifurcation and the latter lattice instability exists, the relationship between the two could be linked by some appropriate procedures, for instance, introduction of the atomic level elastic constants [13] into the macroscopic bifurcation conditions or focus on the soft mode phonon in the relatively long-wavelength region. It is expected that some low-frequency mode should correspond to the bifurcation condition under the uniform homogeneous deformation assumption. Moreover, softening in the short-wavelength region makes it possible to predict an undiscoverable instability by the former. This is the main purpose of the present paper.

A nonequilibrium transition process during phase transformation might principally be driven by the enhanced fluctuation effects. The complex dynamics in the collective system of atoms also brings about instabilities and symmetry breaking leading to reformation of structures on macroscopic scales [14]. Therefore, a secondary aim of this paper is to re-evaluate the extended bifurcation conditions using the atomistic elastic constants defined from the fluctuation formula by Ray and co-workers [15−17]. A quantitative contribution of the fluctuation term to the whole elastic constants is manifested. Then, trace maps of a group of atoms existing in some specific atomic planes during the unstable transformation stage are explored from viewing the spatio-temporal pattern formation.

2. Phase Transformation

2.1 Irreversible Change of Cell

A uniaxial tension is applied at the constant temperature of 900 K to a molecular dynamics simulation cell of 12[100] × 12[010] × 12[001] under the ($H\boldsymbol{\sigma}N$) ensemble [8, 9], where H is the system enthalpy, $\boldsymbol{\sigma}$ the applied stress tensor and N the number of α-iron (α-Fe) atoms in the cell. The potential used here is the embedded atom type many body potential proposed by Finnis and co-workers [18, 19]. A position vector $\boldsymbol{r}^{a}$ of the arbitrary atom α in the cell is expressed as $\boldsymbol{r}^{\alpha} = \boldsymbol{h}\boldsymbol{a}^{\alpha}$, where $\boldsymbol{h}$ is the shape matrix of the cell consisting of three transverse base vectors which characterize the cell shape and $\boldsymbol{a}^{\alpha}$ the normalized position vector of atom α. Thus, $\boldsymbol{h}$ has only three diagonal components before loading for the case of the orthogonal cubic lattice. Each stress increment must be taken small enough to stabilize the coupled vibrations of both the unit cell itself and atoms inside the cell according to two equations of motion under the $(H\sigma N)$ ensemble [8]. Two kinds of extraordinary changes of cell shape are observed just after the stress reached 4.5 GPa as shown in Fig. 2.1.

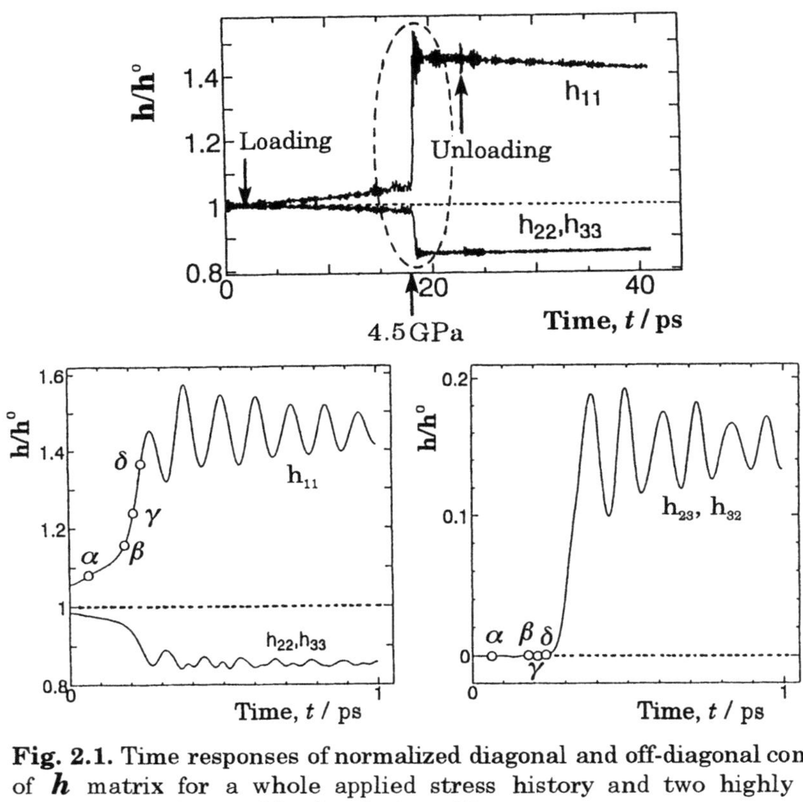

Fig. 2.1. Time responses of normalized diagonal and off-diagonal components of $\boldsymbol{h}$ matrix for a whole applied stress history and two highly resolved responses in the neighborhood of 4.5 GPa.

The two lower figures in Fig.2.1 below the whole applied stress history diagram show the highly resolved responses of diagonal and off-diagonal components of shape matrix $\boldsymbol{h}$ in the neighborhood of 4.5 GPa. Extreme elongation in the tensile direction near the α point and then shearing deformation at the cross-section perpendicular to the tensile direction at the δ point are observed from the lower two figures. The latter deformation can be confirmed by the nonzero off-diagonal component h_{23} and its counterpart h_{32}. From the detailed geometric considerations, as shown in Fig. 2.2, it is concluded that the transformed lattice structure is still the bcc but with the tensile direction of [110], i.e. different from the initial direction of [100]. Therefore, a new cross-section perpendicular to the tensile direction becomes the rhombic structure of the (110) plane in the rectangular coordinate system. Further applied tension after the homogeneous shearing did not produce the hcp structure due to atom shuffling mechanism [3, 8] because of the rather ill-behaved potential employed here for the phase transition [20].

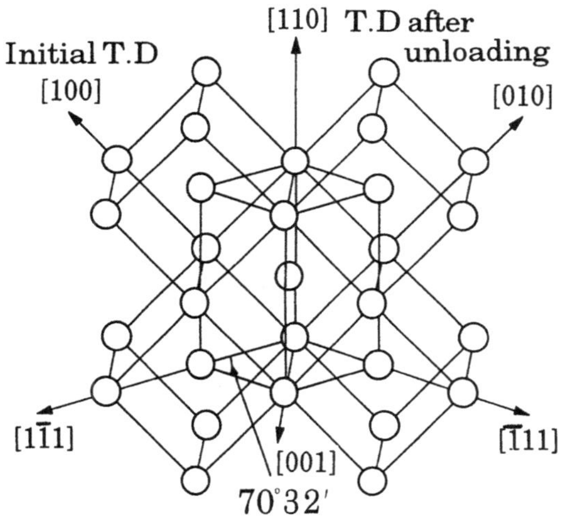

Fig. 2.2. Initial and transformed lattice structures.

2.2 Bifurcations

Wang and Yip have proposed bifurcation conditions under the finite deformation, extending the conventional Born and Huang [21] conditions derived from the positive definiteness of the ordinary fourth-order elastic constants.

$$
\begin{aligned}
B_T &= (C_{11} - \sigma_{xx})(C_{22} + C_{23}) - 2(C_{12} + \sigma_{xx})^2 > 0, \\
G' &= C_{22} - C_{23} > 0, \\
G_1 &= C_{44} > 0, \\
G_2 &= C_{55} - \frac{\sigma_{xx}}{2} > 0
\end{aligned}
\tag{2.1}
$$

The above inequalities with the Voigt notation [13] are evaluated using the atomic level elastic constants defined as

$$C_{ijkl} = \frac{1}{V} \sum_{\alpha} \sum_{\beta \neq \alpha} \Phi_{ik}(\alpha;\beta) r_j(\alpha) r_l(\beta) \quad , \tag{2.2}$$

where the force constants Φ_{ij} are the second derivatives of the interatomic potential and $r_i(\alpha)$ the i-th component of a position vector of atom α. V is the total volume of the cell. A typical pair of these conditions in (2.1) are examined for the relaxation period after the stress has been raised to 4.5 GPa in Fig. 2.3, in which they are found to cross the zero level almost at the above-mentioned points of α and δ, respectively. Loss of the shear modulus of the cross-section perpendicular to the tensile direction, namely, breaking the third inequality of (2.1) with the atomistically estimated elastic constants exactly corresponds to the start of homogenous shearing deformation mentioned previously.

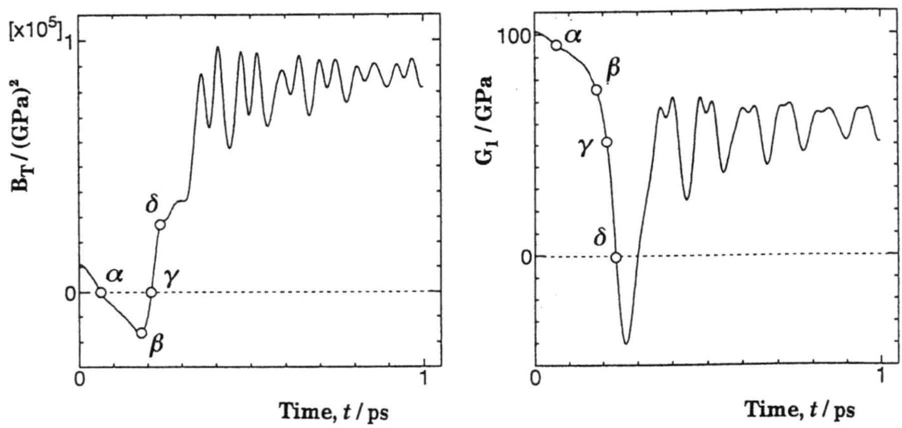

(a) B_T in (2.1) (b) G_l in (2.1)

Fig. 2.3. Macroscopic bifurcation conditions of (2.1) using atomistic elastic constants obtained from the force constants and atomic configuration.

2.3 Phonon Calculations

Let us define a collective velocity $\bar{\boldsymbol{v}}$ at wave vector $\boldsymbol{k}$ under the plane wave approximation as follows [22].

$$\bar{\boldsymbol{v}}(\boldsymbol{k}, t) = \sum_{\alpha} \boldsymbol{v}^{\alpha}(t) e^{i\boldsymbol{k} \cdot \boldsymbol{r}^{\alpha}(t)} \quad , \tag{2.3}$$

where $v^{\alpha}(t)$ is the instantaneous velocity vector of atom α. Phonon spectra density $G_{ii}(\boldsymbol{k},t)$ can be obtained from the correlation function of the collective velocity of (2.3) according to the Wiener−Khinchine theorem.

$$G_{ii}(\boldsymbol{k},t) = \frac{1}{2\pi}\int_{-\infty}^{\infty}\left\langle \bar{v}_i(\boldsymbol{k},t)\bar{v}_i(-\boldsymbol{k},0)\right\rangle \exp(i\,\omega t)dt\,, \qquad (2.4)$$

where <A> denotes the average of a physical value A over a reasonable period of time. G_{ii} is the longitudinal (L) mode if the i-th direction is parallel to the propagation direction k_i, while G_{ii} is the transverse mode (T_1 or T_2) mode if the i-th direction is orthogonal to the $\boldsymbol{k}_i$. The trajectories are integrated for 20 ps after switching the ($H\sigma N$) ensemble into the (NVE) microcanonical ensemble, where E is the total energy. Using spectra distribution obtained by the maximum entropy method [23], phonon dispersion curves are plotted along the prescribed wave vector. The stress dependency of phonon dispersion curves is first studied for deformed lattice structures at some stable stress states. Therefore, the exact change of Brillouin zone for the bct structure is considered here.

Two typical soft mode phonons are summarized in Fig. 2.4. Four Greek symbols in Fig. 2.4 are referred to the four marked instantaneous stages in Fig. 2.1. In the long-wavelength region of the transverse mode (T_1) at the Γ-H propagation direction, the acoustic soft mode phonon is found at the δ point. Namely, it implies that the stationary wave over the whole cell is possible, which is completely analogous over the scale to breaking the third inequality of (2.1) at the δ point.

Another soft mode phonon, zone boundary softening in the short-wavelength region is seen in Fig. 2.4b. This type of soft mode is interpreted as, for example, a change of Brillouin zone due to atom shuffling. The present soft mode phonon analyses also show the possibility of the sequence of instability [3], namely, Burgers' transition mechanism from bcc to hcp: homogeneous shearing deformation at the δ point and then atom shuffling.

3. Fluctuation Effects

3.1 Bifurcations

Ray and co-workers [15−17] have proposed atomistic elastic constants from the fluctuation formula, using the same notation as Parrinello and Rahman [8, 9].

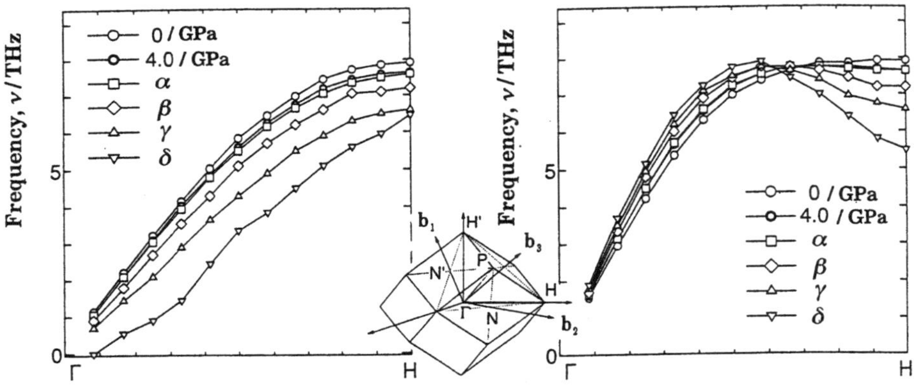

(a) Transverse wave (T_1) (b) Longitudinal wave (L)

Fig. 2.4. Two typical soft mode phonon dispersion curves along the Γ-H propagation direction at several applied stress stages.

$$
\begin{aligned}
V(\boldsymbol{X})(\boldsymbol{h}^{(o)})^{-1}_{ip}(\boldsymbol{h}^{(o)})^{-1}_{jq}(\boldsymbol{h}^{(o)})^{-1}_{kr}(\boldsymbol{h}^{(o)})^{-1}_{ls}[C_{pqrs}] \\
= -\frac{4}{k_B T}\delta(m_{ij}m_{kl}) + 2Nk_BT((\boldsymbol{G})^{-1}_{li}(\boldsymbol{G})^{-1}_{jk} \\
+(\boldsymbol{G})^{-1}_{ki}(\boldsymbol{G})^{-1}_{jl}) + \left\langle \sum_{\alpha}\sum_{\beta>\alpha}\left(\frac{\partial^2\Phi(r^{\alpha\beta})}{\partial(r^{\alpha\beta})^2} - \frac{1}{r^{\alpha\beta}}\frac{\partial\Phi(r^{\alpha\beta})}{\partial r^{\alpha\beta}}\right)\right. \\
\left. \times\frac{a_i^{\alpha\beta}a_j^{\alpha\beta}a_k^{\alpha\beta}a_l^{\alpha\beta}}{(r^{\alpha\beta})^2}\right\rangle . \qquad (3.1)
\end{aligned}
$$

$V(\boldsymbol{X})$ is the total volume of the cell measured at the reference framework and $\boldsymbol{h}$ is the shape matrix mentioned in the previous section. $\boldsymbol{G}$ is the metric tensor, $\boldsymbol{h}^T\boldsymbol{h}$, and $\boldsymbol{m}$ is a newly defined stress tensor [15]. The first term of the right-hand side of (3.1) indicates the fluctuation contribution, the second the kinetic part and the last one the potential part into which (2.2) is reformulated for the pair-wise potential. These terms are examined in sections near the δ point in Fig. 3.1, in which the boosted fluctuation contribution makes some components of the total elastic constants negative, as C_{33} shown here.

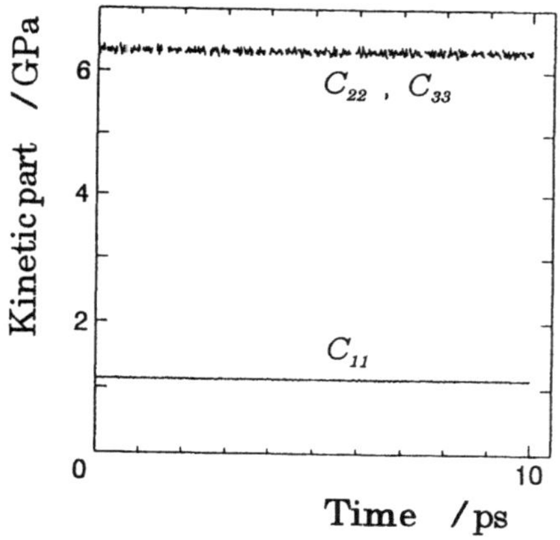

(a) kinetic part

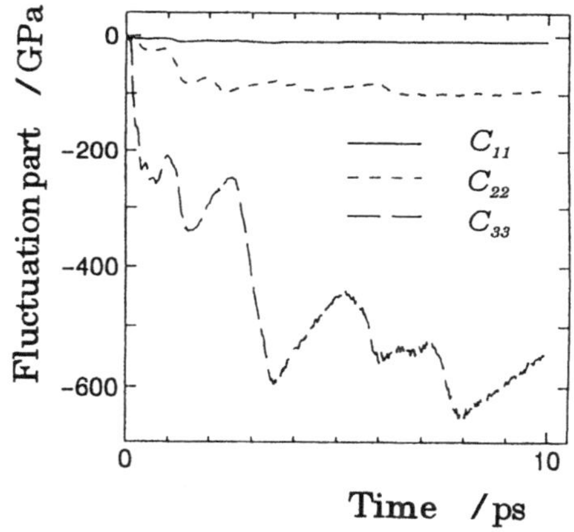

(b) fluctuation part

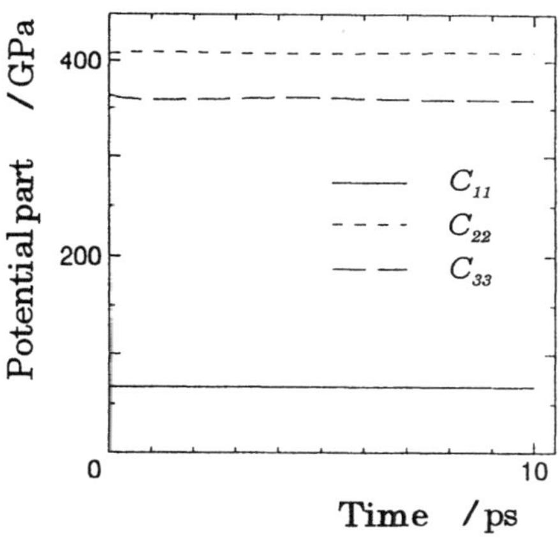

(c) potential part

Fig. 3.1. Fluctuation effects by Ray's elastic constants near the δ point.

3.2 Pattern Formations

Figure 3.2 shows two snapshots of the collective system of atoms, the first step (displaying two kinds of layers by ◇ and ◆) is just after increasing the stress to 4.5 GPa and a step (○ and ●) after 10ps relaxation at 4.5 GPa. Figure 3.2a b are a (100) atomic plane perpendicular to the tensile direction and a (110) plane in which a family of slip directions of the bcc structure, <111> , originally in-plane exists, respectively. The local shear deformation pattern according to the metallurgical slip system could be the basic mechanism in order to reform from the orthogonal cross-section to the rhombic. The preferred slip flow may lead to the nucleation of some curious patterns, such as a vortex.

One of the most challenging topics of phase transformation is currently the nucleation of an embryo and its growth mechanism. Some experimental reports have already been published [24, 25], in which the phase transformation begins from a strain concentration portion such as at an interface to an inclusion, a grain boundary, a surface step and a dislocation tangle [25]. The nucleation of stress-induced phase transformation has also been observed at the intersection of microbands [26]. Interaction between a specific frozen phonon mode and lattice defect may explain this [24]. However, the general physical mechanism of nucleation is still unclear. It is expected that a study of the nonlinear spatio-temporal pattern formation produced by the interaction of a phonon mode and a metallurgical slip system could be incorporated into the fundamental mechanism of nucleation.

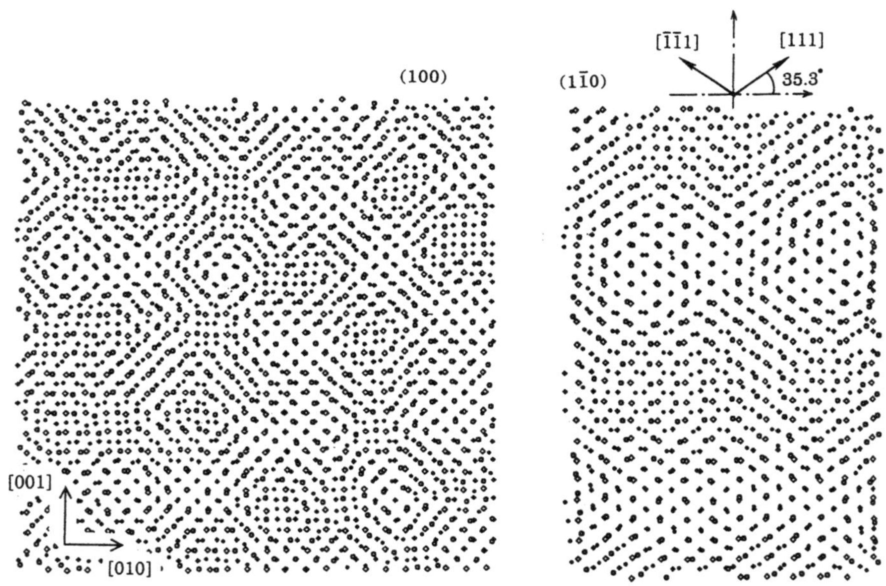

(a) (100) atomic plane

(b) (110) atomic plane

Fig. 3.2. Spatio-temporal pattern formations obtained during extremely fluctuated relaxation process towards the phase transformation.

4. Conclusions

The scale-dependent relation between the macroscopic bifurcation conditions and a soft mode phonon in the stress-induced phase transformation has been studied by molecular dynamics simulations. It is noted that loss of the shear modulus of the cross-section perpendicular to the tensile direction, namely, breaking the third inequality of (2.1) exactly corresponds to the start of reforming to the rhombic shape. Moreover, one of the soft mode phonons, the acoustic soft mode in the long-wavelength region is found to be completely analogous over the scale to breaking that inequality. Fluctuation effects on the atomic level elastic constants are quantified and then found to be more active when losing the shear modulus. Dissipation coupled to the growing fluctuation might bring some spatio-temporal pattern formation with the basic flow pattern governed by the metallurgical slip system.

Acknowledgements

The studies concerning the molecular dynamics simulations were performed by Mr. J. Toryu, a graduate student of Kobe Univ. (currently Kawasaki Steel Corp.). The author thanks Prof. Y. Tomita of Kobe Univ. for helpful discussions. This work was partially supported by a grant of the Minister of Education, Science and Culture of Japan.

References

1. L.D. Landau, E.M. Lifshitz: *Statistical Physics*, (Pergamon Press, Oxford, 1980).
2. Y.C. Fung: *Foundations of Solid Mechanics*, (Prentice-Hall Inc., New Jersey, 1965).
3. K.S.Cheung, R.J.Harrison, S.Yip: J. Appl. Phys. **71**, 4009−4014 (1992).
4. J. Wang, S. Yip, S.R. Phillpot, D. Wolf: Phys. Rev. Lett. **71**, 4182−4185 (1993).
5. K. Mizushima, S. Yip, E. Kaxiras: Phys. Rev. B **50**, 14952−14959 (1994).
6. M. tang, S. Yip: Phys. Rev. Lett. **75**, 2738−2741 (1995).
7. J. Wang, J. Li, S. Yip, S. Phillpot, D. Wolf: Phys. Rev. B **52**, 12627−12635 (1995).

8. M. Parrinello, A. Rahman: Phys. Rev. Lett. **45**, 1196–1199 (1980).
9. M. Parrinello, A. Rahman: J. Apply. Phys. **52**, 7182–7190 (1981).
10. W.G. Burgers: Physica **77**, 533 (1953).
11. F. Milstein: Phys. Rev. B **3**, 1130–1141 (1971).
12. N.W. Ashcroft, N.D. Mermin: *Solid state physics*, (Saunders College, 1976).
13. D.C. Wallace: *Thermodynamics of crystals*, (Wiley, New York, 1972).
14. D. Walgaef: *Spatio-Temporal Pattern Formation*, (Springer 1996).
15. J.R. Ray, A. Rahman: J. Chem. Phys. **80**, 4423–4428 (1984).
16. J.R. Ray, M.C. Moody, A. Rahman: Phy. Rev. B **32**, 733–735 (1985).
17. J.R. Ray: *Elastic constants and statistical ensembles in molecular dynamics*, (North-Holland, Amsterdam, 1988).
18. M.W. Finnis, J.E. Sinclair: Phil. Mag. A **50**, 45–55 (1984).
19. G.J. Ackland, G. Tichy, V. Vitek, M.W. Finnis: Phil. Mag. A **56**, 735–756 (1987).
20. M. Ludwig, P. Gumbsch: Modelling. Simul. Mater. Sci. Eng. **3**, 533–542 (1995).
21. M. Born, K. Huang: *Dynamical theory of crystal lattice*, (Clarendon, Oxford, 1956).
22. X.Q. Wang, G.L. Chiorotti, F. Ercolessi, E. Tosatti: Phys. Rev. B 38, 8131–8134 (1988).
23. W.H. Press, S.A. Teukolsky, W.T. Vetterling, B.P. Flannery: *Numerical recipes in FORTRAN*, (Cambridge univ., 1992).
24. Y. Yamada: Proc. ICMT, Japan 1986, pp.89–94.
25. T. Saburi, S. Nenno: Proc. ICMT, Japan 1986, pp.671–678.
26. G.B. Olson, M. Cohen: Metall. Trans. A **6A**, 791–795 (1975).

Propagation Dynamics of Fractal Cracks

Igor L. Maksimov

Nizhny Novgorod University, N. Novgorod, 603600 Russia

Summary. The global energetics of the fractal crack is studied taking into account the stress-field singularity weakening near the self-affine crack. The governing power exponents are determined for the generalized self-affine plastic zone condition. The griffith stability condition is discussed, and the fractal crack runaway dynamics is described.

Key Words. Crack propagation, fractal fracture, plastic zone, runaway dynamics

1. Introduction

Since the invention of the fractal concept, much attention has been paid to apply this method to describe the fracture phenomenon in both brittle and ductile materials (see e.g. [1–3]). It has been ascertained that the conventional linear mechanics assumption of the smooth crack surface is valid only on the macroscopical level. Within the mesoscopic crack scale

$$l_0 \ll r \ll R_c$$

(here, l_0 is of the order of the micrograin size ($l_0 \simeq 1\ \mu m$) and $R_c \simeq 10^{-2}\ m$ is the crack-surface self-affine correlation length) the crack surface becomes quite rough. This circumstance requires the use of an adequate mathematical description in terms of a self-affine (or, simply, fractal) crack.

In the present paper the scaling exponents for the global quantities controlling the crack energetics (energy release rate, surface energy, plastic work) will be derived for the self-affine cracks. The dynamics of the fractal crack propagation will be described. The monotonous crack acceleration is predicted for the first time during the mesoscopic stage of crack advance.

2. Fractal Crack Geometry and Stress Distribution

The governing model for conventional fracture theory is a quasi-elliptical crack with macroscopically smooth faces [4]. The problem of the stress and strain distribution near the only crack singular point – the crack tip – can be solved by means of linear elastic mechanics (LEM) methods. However, both experimental and numerical data show that the surface profile of the crack can be of a fractal nature. Therefore, the modification of the LEM concept to describe the fractal crack energetics and growth is an as yet unsolved problem in fracture theory.

2.1 Geometry of a Self-Affine Crack

Dimensional characteristics of a self-affine crack shown in Figure 1 are governed by the Hurst exponent $0 < H \leq 1$, which is a measure of the surface roughness. For example, the box-counting dimension D_B is related to the topological dimension D of the system as

$$D_B = D - H. \tag{1}$$

The true crack length (surface) a_t scales to the crack projection length a by means of the trail latent dimension $D_1 = min\{\frac{D-1}{H}, D\}$:

$$\ln a_t \sim D_1 \ln a. \tag{2}$$

As is seen from Fig.1, the normal direction $\mathbf{n}$ can be defined for any local coordinate system. In what follows we concentrate on the two-dimensional case $D = 2$ assuming the sample to be a slab of unit thickness $h = 1$ under uniaxial tension p (see Fig. 1).

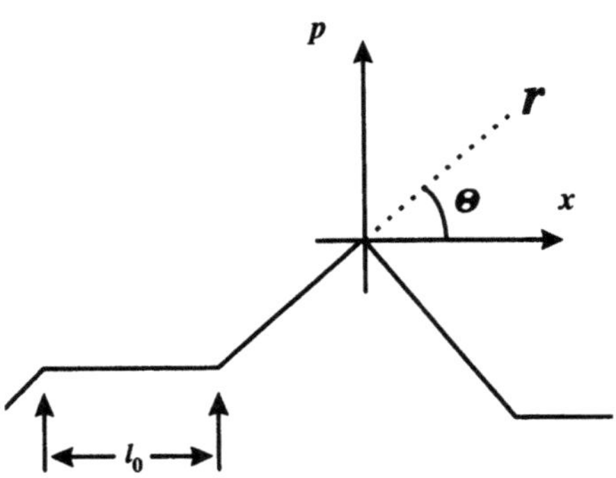

Fig. 1. The geometry of fractal crack.

2.2 Stress Singularity Near a Self-Affine Crack

Geometrical peculiarities of the crack trajectory significantly affect the stress distribution near the crack. The spatial stress distribution for a specific crack shape is an extremely complicated problem, therefore one should resort to a statistical description of the stress field. By employing the superposition principle it is possible to obtain the stress-concentration law near the fractal crack taking into account the distribution function $Q(-x) \propto x^{\epsilon(1-H)}$ of singular points within the interval (-x,0):

$$\sigma_{ij}(r,\theta) \propto \phi_{ij}(\theta) \int_{-R_c}^{0} \frac{Q(-x)}{R(x,r)^{3/2}} dx \propto K_{ef} r^{-\alpha}. \tag{3}$$

The effective stress intensity factor K_{ef} is a linear combination of the K_I and K_{II} components (since the tension fracture mode is intermixed with the shearing one) [5]. The corresponding displacement field is described by

$$u_i = \tilde{\phi}_i(\theta) \frac{K_{ef}}{E} r^{1-\alpha}. \tag{4}$$

The nondimensional functions $\phi_{ij}(\theta)$ and $\tilde{\phi}_i(\theta)$ in (3), (4) describe the angular dependence of the stress tensor and displacement vector components, respectively. It is seen from (3) that the stress-singularity power exponent $\alpha = 0.5 - \epsilon(1 - H)$ is smaller than that for a smooth crack ($\alpha_s = 0.5$). Therefore, we conclude that the stress-field near the fractal crack becomes less singular. Taking into account the data ($\epsilon \simeq 2$) reported by [6] we conclude that the roughness-induced suppression of the singularity takes place at $H < H^* = 1 - (2\epsilon)^{-1} \simeq 0.75$.

2.3 Geometry of a Self-Affine Plastic Zone

The singularity of the fractal crack stress field provides the condition for the dislocation emission and consequently for the plastic strain localization near the crack tip(s). It is usually assumed (or derived) that the shape of a plastic zone is invariant for any crack length (see Fig. 2) and its size

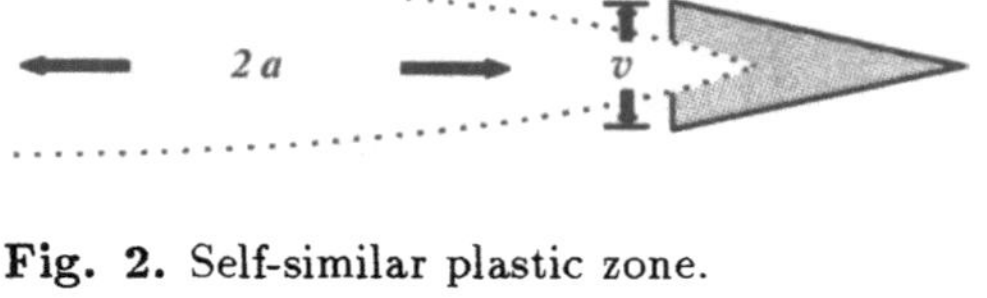

Fig. 2. Self-similar plastic zone.

(projection length) d is controlled by the ratio of the external stress p and the plastic yield stress σ_Y:

$$\frac{d}{a} = \Psi(\frac{p}{\sigma_Y}) \tag{5}$$

where $\Psi(z)$ is the function found from the condition for the stress-singularity saturation [7]. The relation (5) uniquely determines the plastic zone (PZ) volume V^{pl} and surface area A^{pl} as a function of the actual crack length a. To describe the generalized self-similarity of the plastic zone (GSSPZ) a modified relation

$$\ln d = \ln \Psi^*(p/\sigma_Y, l_0) + r \ln a \tag{6}$$

has to be used in the case of a complex PZ morphology. For the limiting case $r = 0$ the expression (6) corresponds to the Barenblatt model [8]; for the case $r = 0$ (6) reduces to the Dugdale–Leonov–Panasyuk model [4].

As yet, there is no closed solution for the spatial distribution of the plastic deformation in the general case of a concrete fractal crack. However, recent experimental and numerical studies of this problem show that plastic zone(s) near the singularity(ties) can also acquire the fractal shape. Therefore, the concept of the self-similar plastic zone (SSPZ) should be modified in accordance with the self-affine plastic zone (SAPZ) definition (see Fig.3). Accordingly, the plastic zone volume per unit thickness V^{pl} is related to the plastic zone size d as

$$\ln V^{pl} \propto D_2^{pl} \ln d \tag{7}$$

where $D_2^{pl} \simeq D$ is a volume fractal dimension. The plastic zone surface area per unit thickness A^{pl} is related to the plastic zone size d as

$$\ln A^{pl} \propto D_1^{pl} \ln d \tag{8}$$

where $1 < D_1^{pl} < D$ is a surface fractal dimension, governed by the roughness exponent H.

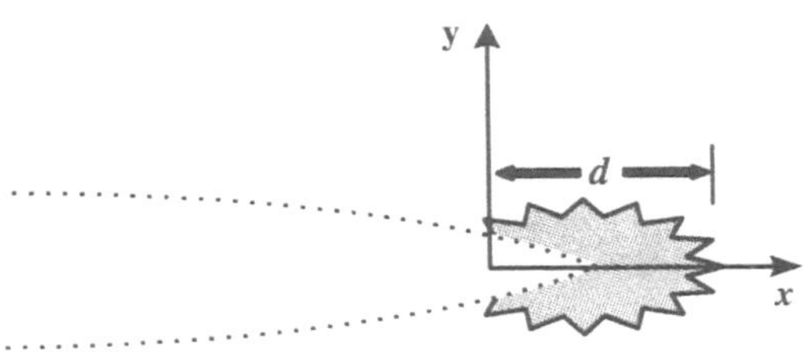

Fig. 3. Self-affine plastic zone.

By employing (5) or (6) it is possible to express the V^{pl} and V^{pl} in terms of crack length a for any specific PZ shape.

3. Global Energetics of Self-Affine Cracks

Fractal features of the crack geometry result in a modified crack-length dependence of the fracture energies. Therefore, the fractal-crack stability condition requires a more detailed study. The self-affine crack energy characteristics will be evaluated below.

3.1 Elastic Energy Release Rate

The smoothing of the stress-field singularity near the fractal crack leads to a higher elastic energy release rate in comparison with the macroscopically smooth crack. The elastic energy stored within the long-range power-law tail of the stress field for our geometry (see Fig. 1) is evaluated as [4]

$$U_{el} = \int\int \sigma_{nj}(r,\theta)u_j(r,\theta)d^2r \simeq \frac{p^2}{E}\frac{l_0^2}{3-2\alpha}\langle\phi_{nj}\tilde{\phi_j}\rangle[\frac{a}{l_0}]^{(3-2\alpha)} \equiv \Phi a^{\beta}. \qquad (9)$$

In (9) we introduced the average over the angle θ . The elastic energy power exponent $\beta = 2 + \epsilon(1-H)$ for a fractal crack exceeds that for the smooth crack $\beta_s = 2$ [8]. This feature, as shown below, qualitatively modifies the fractal crack propagation dynamics. From (9) it follows immediately that the power exponent for the elastic energy release rate $G = \partial U_{el}/\partial a \propto a^{\beta-1}$ is equal to $\beta - 1$. Note that for $\epsilon = 2$ an exponent β is related to the box-counting dimension D_B : $\beta = 2D_B$.

3.2 Crack Surface Energy

The crack-length dependence of the crack surface energy U_S is governed by the trail latent dimension D_1 which takes a value $D_1 \simeq 1$ for a smooth crack surface ($H \to 1$), and $D_1 = 2$ for a sufficiently rough crack surface ($H < 0.5$). Therefore, introducing the surface tension γ one finds:

$$U_S(a) \simeq 2\gamma l_0[\frac{a}{l_0}]^{D_1} \equiv \Gamma a^{D_1}. \qquad (10)$$

It is seen from (10) that the fractal crack surface energy may significantly (up to 1–2 orders of magnitude) exceed the value for a smooth crack surface [2].

3.3 Dissipation Power Exponents for Quasi-Brittle Fracture

The energy dissipated during plastic deformation is usually estimated in terms of plastic work [7 , 9]

$$W_{pl} = \int_0^d [\sigma_{yy}\, u_y]|_{u_y=v(x)} dx \simeq G_{pl} d^2 \tag{11}$$

where $u_y = v(x)$ is the plastic zone profile. For a GSSPZ case ($d \propto a^r$) the expression for W_{pl} takes a form:

$$W_{pl} = G_{pl} a^q \tag{12}$$

with a plastic work power exponent $q = 2r \leq 2$. A similar relation holds for a self-affine plastic zone case for which $q = rD_2^{pl} \leq 2$.

3.4 Griffith Stability Condition

Let us now discuss the fractal crack stability. According to Griffith [8], the crack is stable, provided

$$G_{el} \leq \frac{\partial [U_S + W_{pl}]}{\partial a}$$

or, substituting (9), (10) and (12), one obtains

$$\beta \Phi a^\beta \leq D_1 \Gamma a^{D_1} + q G_{pl} a^q. \tag{13}$$

Since $\beta > max\{D_1, q\}$ we conclude that a sufficiently large crack of size $a > a_{cr}$ becomes unstable and tends to accelerate monotonously. The critical crack length a_{cr} is determined from (13).

4. Crack Runaway Dynamics

4.1 Dynamic Energy Balance

cribe the crack propagation dynamics one should examine the energy balance taking into account the crack kinetic energy

$$T(a, \dot{a}) = \int\int \Delta m(x,y) \dot{\mathbf{u}}^2(x,y) d^2 r. \tag{14}$$

The latter can be evaluated using a Mott scaling scheme [8]. Since the stress distribution is of the power-law type, the dominant contribution to the integral (14) is provided by the far-from-the-crack region of size a that is not sensitive to the details of the crack profile. Therefore, the conventional Mott estimate for the kinetic energy can be used:

$$T(a,\dot{a}) = 0.5\mu a^2 \dot{a}^2 \tag{15}$$

where the crack "inertia" coefficient μ is calculated from the quasi-static stress field.

4.2 Supercritical Crack Runaway Phenomenon

Introducing (15) into the energy-conservation law

$$\Delta U_{el}(a) = T(a,\dot{a}) + \Delta U_S(a) + \Delta W_{pl}(a) \tag{16}$$

where $\Delta F(a) = F(a) - F(a_{cr})$ one finds that the crack-velocity is dependent on the crack length a:

$$\dot{a} = a^{-1}\sqrt{\frac{2}{\mu}}\{\Phi[a^{\beta} - a_{cr}^{\beta}] - \Gamma[a^{D_1} - a_{cr}^{D_1}] - G_{pl}[a^q - a_{cr}^q]\}^{1/2} \tag{17}$$

From (17) it follows that near the Griffith instability threshold $(a-a_{cr})/a_{cr} \ll 1$, the crack velocity is controlled by a lowest dissipation-power exponent $s = min\{q, D_1\}$:

$$\dot{a} \propto (a - a_{cr})^{s/2}$$

which leads to a power-law crack advance dynamics at the initial stage:

$$a(t) - a_{cr} \propto t^{2/(2-s)}$$

(note that for the case s = 2 crack elongation is exponentially fast: $a(t) - a_{cr} \propto \exp(t/t_0)$). During the later stage $a(t) \geq a_{cr}$ the crack velocity is controlled by the energy release exponent $\beta \geq 2$:

$$\dot{a} \simeq \sqrt{\frac{2\Phi}{\mu}} \quad a^{(\beta-2)/2} \tag{18}$$

which results in a runaway crack-advance dynamics

$$a(t) = [\frac{(4-\beta)^2\Phi}{2\mu}]^{1/(4-\beta)} \quad t^{2/(4-\beta)}. \tag{19}$$

Note that the crack velocity monotonically increases with time

$$\dot{a}(t) \propto t^{(\beta-2)/(4-\beta)}$$

due to nonzero acceleration $\ddot{a}(t) > 0$ in the course of the crack-runawa y stage

$$\ddot{a}(t) \propto \frac{(\beta-2)}{(4-\beta)^2} \quad t^{(2\beta-6)/(4-\beta)}.$$

We should take into account that the monotonous acceleration stage will eventually be terminated when a crack enters the region of macroscopically smooth surface morphology ($a(t) > R_c$). In the final stage, the crack propagation is governed by the LEM equations, which eventually result in the crack velocity saturation at the terminal value $\dot{a}(t) \simeq (\dot{a})_T$ when $t \to \infty$. The cross-over time t_c for which $a(t_c) = R_c$ is estimated as

$$t_c \propto R_c^{(4-\beta)/2} \quad .$$

Crack advance dynamics is shown schematically in Figure 4.

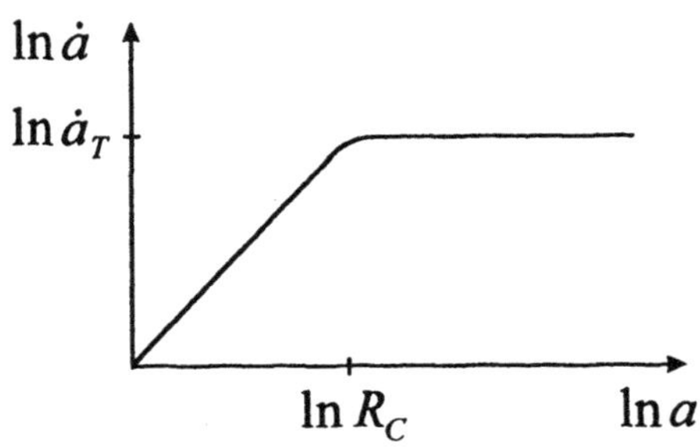

Fig. 4. Fractal crack advance dynamics.

5. Discussion

The results obtained are applicable to describe quasibrittle fracture of fractal cracks, since in this case the elastic energy release rate dominates over dissipation processes [9]. Note that the constant-load condition was assumed in our study. In the more general case of decreasing load: $dp/dt < 0$ the crack dynamics may be quite different [10].

The opposite case of stable crack growth typical for fracture in metals [2] requires a more detailed description. We wish to note that in the latter case the crack advance condition is controlled mainly by the local energetics near the crack tip(s). This needs further study taking into account the specifics of grain boundaries, morphology, etc.

Energy dissipation taking place during quasibrittle fracture leads to local heating near the crack. The latter may significantly affect the plastic yield stress and, in accordance with (5), (6) and (12), may appreciably modify the fractal crack dynamics. Therefore, a study of a non-isothermal propagation of a self-affine crack would be of significant interest.

6. Conclusions

i) Power exponents for the energy release rate, surface energy and the plastic work are related to the dimensional characteristics of self-affine cracks. The energy release rate exponent is shown to be greater than that for dissipation energies and is found to govern fractal crack dynamics.

ii) The runaway dynamics of the fractal crack propagation is predicted in the course of quasibrittle fracture. It represents a monotonous crack acceleration during the fractal stage of crack advance. The cross-over time to the terminal propagation velocity is found.

Acknowledgement

The author wishes to thank Prof H. Kitagawa and Dr A. Nakatani for valuable discussions on the problem. I also appreciate the JSPS support in accordance with the JSPS visiting researcher program.

References

1. H. Takayasu: *Fractals in the physical sciences,* (Manchester University Press, New York, 1990).
2. C.W. Lung: *Fractals and the fracture of cracked metal. In Fractals in physics,* L. Pietroniero and E. Tosatti (eds), (North-Holland, Amsterdam, 1986) pp.189–192.
3. E. Louis, F. Guinea: *The fractal nature of fracture*, Ibid. pp.177–180.
4. G.P. Cherepanov: *Mechanics of brittle fracture,* (McGraw-Hill, New York, 1979).
5. A.S. Balankin: Phil.Mag.Lett. **74**,415 (1996).
6. V.V. Mourzenko, J.-F. Thovert, P.M. Adler: Phys.Rev. E**53**,5606 (1996).
7. K. Hellan: *Introduction to fracture mechanics,* (McGraw-Hill, New York, 1984).
8. B. Lawn: *Fracture of brittle solids,* (Cambridge University Press, New York, 1993).
9. I.L. Maksimov, N.G. Golubeva, G.M.Maksimova, V.S. Metrikin, V.N. Smirnova, J.V. Svirina: Mater. Sci. Eng. A **176**, 309 (1994).
10. J. Schmittbuhl, K.J. Måløy: Phys. Rev. Lett. **78**, 3888 (1997).

Criteria for Nucleation of a Dislocation and a Cleavage Crack in a Nickel Single Crystal Based on Molecular Dynamics

Takayuki Kitamura, Kisaragi Yashiro and Ryuichi Ohtani

Department of Engineering Physics and Mechanics, Graduate School of Engineering, Kyoto University, Kyoto 606-8317, Japan

Summary. The microscopic features of deformation and fracture in a nickel single crystal are investigated on the basis of the local strain of the crystal lattice. The process of dislocation nucleation from the surface without the constraint as well as the process of cleavage cracking under tension, with the constraint of transverse deformation, are simulated by molecular dynamics, and it is found that their nucleation criteria are successfully derived by the lattice instability.

Key Words. Local criterion, lattice instability, nucleation, fracture, dislocation, molecular dynamics

1. Introduction

Intense development of components with nano-structure for advanced electronic devices has taken place. Nucleation of a single defect on the atomic scale sometimes brings about fatal malfunction of such devices. Therefore, it becomes important to understand the mechanics and mechanism of defect nucleation on a microscopic scale. Although the fracture of materials has been investigated for many years, there remain many uncertainties with regard to the nucleation process because of experimental difficulties.

The recent rapid progress of computers makes the microscopic simulation of solid structures possible. In particular, fracture processes such as crack growth behavior and deformation in the vicinity of a crack tip, including the characteristic feature of dislocation emission [1], are well reproduced with the molecular dynamics (MD) technique. Although simulations of crack growth are useful for understanding the fracture process of large components, it is doubtful that a pre-existing crack dominates the life of micro-devices. The fracture of devices with carefully fabricated structure might be governed by the nucleation and/or multiplication process of microscopic defects rather than by the process of crack growth. However, few studies have been devoted to defect nucleation [2].

Tensile fracture of metals is mainly classified into two types; brittle fracture without plastic deformation and ductile fracture accompanied by necking of the crystal. Figure 1.1 shows typical fracture processes simulated for a nickel single crystal subjected to a tensile stress along the [001] direction with the transverse constraint (Fig. 1.1a) and without it (Fig. 1.1b) by molecular dynamics [3, 4]. Although the latter is generally observed by experiments in a fcc metal, the former is possible under "strict constraint" of a small crystal. The ductile fracture is brought about by the glide of dislocations on the (111) planes as illustrated in Fig. 1.1b. When the slip is strictly restricted by the constraint of transverse deformation, it brings about cleavage fracture on the (001) plane as shown in Fig. 1.1a. In this study, the nucleation processes of a cleavage crack, as well as a dislocation, are simulated by molecular dynamics, and the local strain conditions at the onsets of their nucleation obtained by molecular dynamics simulation is discussed in terms of the lattice instability criteria in a perfect crystal under uniform deformation.

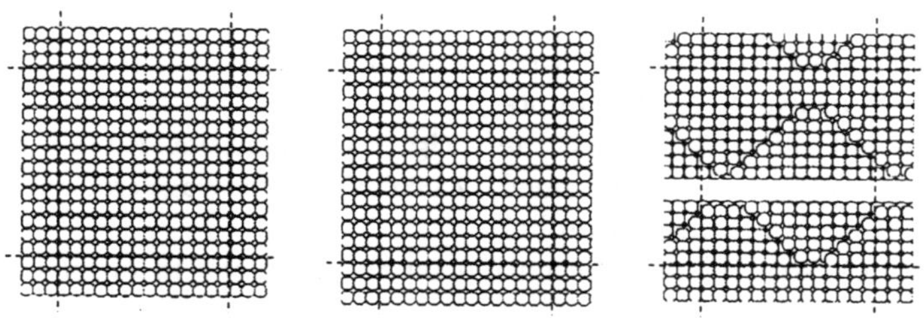

$\varepsilon = 0.10$ $\varepsilon = 0.13$ $\varepsilon = 0.158$

(a) Brittle fracture under tension with constraint of transverse deformation

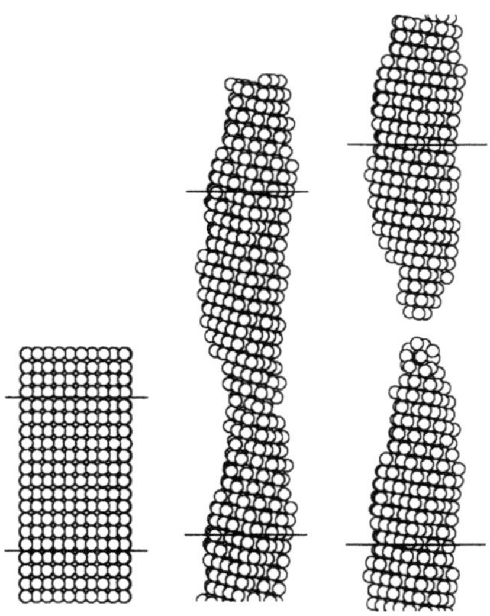

$\varepsilon = 0.10$ $\varepsilon = 1.50$ $\varepsilon = 1.90$

(b) Ductile fracture due to dislocation glide

Fig. 1.1. Typical fracture processes under tension simulated by molecular dynamics.

2. Criterion on the Nucleation of a Cleavage Crack

2.1 Simulation Procedure

Simulation is carried out for a nickel single crystal using a simulation cell illustrated in Fig. 2.1 with 1372 atoms (343 fcc unit cells) with periodic boundary conditions [5]. The perfect crystal is subjected to the tensile strain, $\varepsilon_{zz}^{(\mathrm{global})}$, of 0.001 in a [001] direction. The spacing of all atoms is proportionally expanded in the z-axis during straining in order to avoid the generation of a stress wave. The cell size in the xy-plane is preserved during the simulation, so that the crystal is subjected to the tensile stresses not only on the z-direction but also in the transverse directions. Then, it is relaxed by the molecular dynamics simulation for 1 ps at a temperature of 2.5 K. The procedure of straining/relaxation is repeated until the crystal fractures. If the potential energy does not converge in 1 ps, the molecular dynamics simulation is continued until a stationary state is reached. Emphasis is on the "strict constraint" of deformation on the xy-plane in order to reproduce cleavage cracking in a small fcc (metallic) crystal.

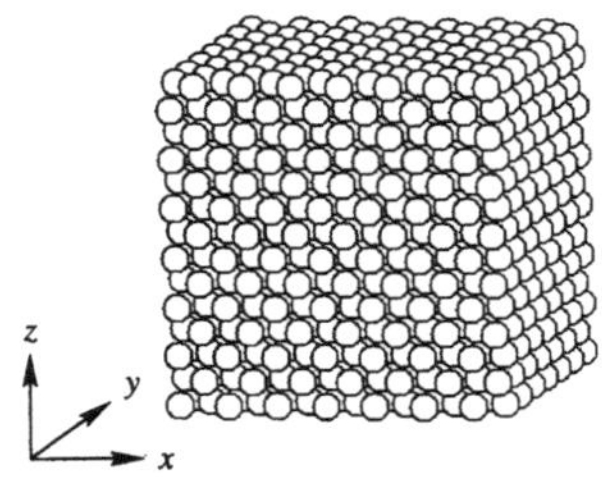

Fig. 2.1. Simulation cell in the calculation of cleavage cracking.

The interatomic force is evaluated by the embedded atom method (EAM) [6], in which the total potential energy, E_{tot}, is formulated by

$$E_{\mathrm{tot}} = \sum_{i=1}^{N} F_i(\rho_i) + \frac{1}{2} \sum_{i=1}^{N} \sum_{\substack{j=1 \\ (j \neq i)}}^{N} \phi_{ij}(r_{ij}) \ , \tag{2.1}$$

where $F_i(\rho_i)$ is the energy for embedding an atomic nucleus at the location with the electron density of ρ_i, ϕ_{ij} is the energy originating from the repulsion of nuclei, and r_{ij} is the distance between nuclei i and j. The functions and constants adopted are described in detail in the original paper [6].

2.2 Simulation Results

The solid line in Fig. 2.2 shows the relationship between the stress, $\sigma_{zz}^{(\mathrm{global})}$, and the strain, $\varepsilon_{zz}^{(\mathrm{global})}$. Here, $\sigma_{zz}^{(\mathrm{global})}$ is evaluated by the average of the stress acting on the each atom, $\sigma_{zz}^{(\mathrm{local})}$, in the crystal. The local stress,

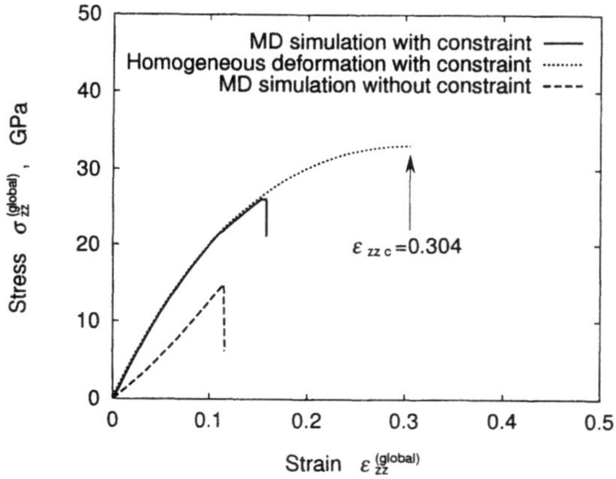

Fig. 2.2. Relationships between global stress and global strain under tension.

$\sigma_{zz}^{(\text{local})}$, is evaluated by [7]

$$\sigma_{\alpha\beta}^{(\text{local})} = \frac{1}{\Omega}\left[-mv_i^{\alpha}v_i^{\beta} + \frac{1}{2}\sum_{j\neq i}\left\{\left(F'(\rho_i) + F'(\rho_j)\right)\rho'(r_{ij}) + \phi'(r_{ij})\right\}\frac{r_{ij}^{\alpha}r_{ij}^{\beta}}{r_{ij}}\right], \tag{2.2}$$

where Ω is the atomic volume, m is the atomic mass, v_i^{α} is the α-component ($\alpha = x$, y or z) of the velocity of atom i, and r_{ij}^{α} is the α-component of the vector from atoms i to j. Plastic deformation does not take place before fracture, and the nonlinearity in the elastic deformation originates from the function of the interatomic potential [3].

At the global strain of $\varepsilon_{zz}^{(\text{global})} = 0.158$, a crack is initiated and then grows along a (001) plane, as illustrated in Fig. 1.1a [3]. Figure 2.3 shows the local strain, $\varepsilon_{zz}^{(\text{local})}$, which is evaluated from the distance between adjacent atoms on the neighboring (001) planes. Before the stress drops at $\varepsilon_{zz}^{(\text{global})} = 0.158$, there is little fluctuation in the local strain, as shown in Fig. 2.3a. In other words, the crystal deforms uniformly. The fluctuation appears in the local strain at the instance when slip takes place at the stress drop, as illustrated in Fig. 2.3b, and the cleavage crack nucleates between Planes 5 and 6 indicated by an arrow, where the local strain is maximal in the crystal (see Fig. 1.1a). This suggests that the nucleation is governed by the concentrated local tensile strain due to slip.

We examined the effect of the simulation cell size on the nucleation of a cleavage crack in a previous paper [5], and this study revealed that the global fracture strain decreased with the increase in the cell size due to the fluctuation of the local strain. However, the cleavage crack nucleates on the (001) plane with the maximum local strain, $\varepsilon_{zz}^{(\text{local})}$, in all cases, and the critical strain was about 0.31, which coincides with the local strain obtained from Fig. 2.3b. This signifies that the local strain is an appropriate candidate for the criterion of the crack nucleation.

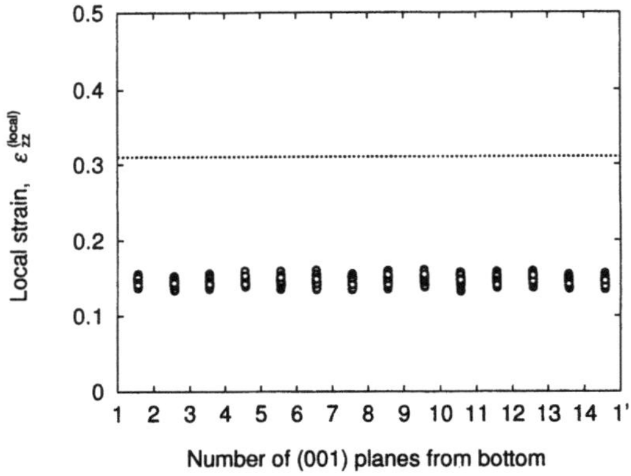

(a) Before stress drop ($\varepsilon = 0.148$)

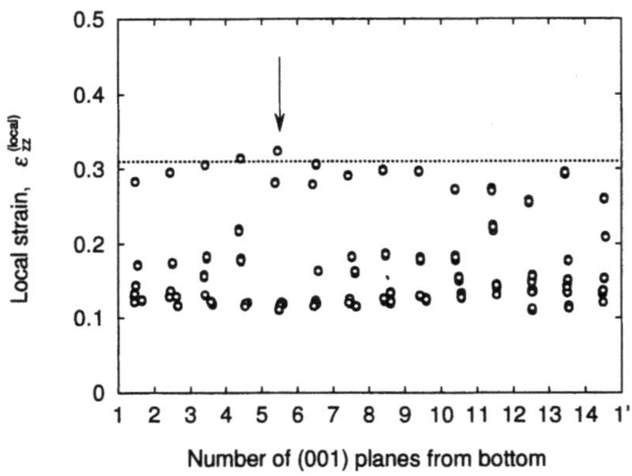

(b) At stress drop ($\varepsilon = 0.158$)

Fig. 2.3. Local tensile strain in the crystal. Abscissa represents number of (001) planes from the bottom of the simulation cell illustrated in Fig. 2.1.

2.3 Local Instability Criterion

In order to analyze the nucleation condition, the tensile process is idealized assuming that all atoms are located on the fcc lattice points and the crystal deforms uniformly. Thus, all fcc unit cells in the crystal deform equally and the deformation is represented by the variables, a_1 to a_6, illustrated in Fig. 2.4. The idealized tensile behavior under the transverse constraint is calculated by increasing the magnitude of a_3 while $a_1 = a_2 = a_0$ and $a_4 = a_5 = a_6 = \pi/2$, where a_0 is the lattice constant at the initial state. The dotted line in Fig. 2.2 indicates the $\sigma_{zz}^{(\mathrm{global})}$–$\varepsilon_{zz}^{(\mathrm{global})}$ relation, which is the same as $\sigma_{zz}^{(\mathrm{local})}$–$\varepsilon_{zz}^{(\mathrm{local})}$ in the idealized homogeneous deformation. The relation coincides very well with the result in the molecular dynamics simulation until the cleavage fracture takes place.

Milstein [8] derived the instability condition of lattice shape and size under homogeneous deformation on the basis of an energy balance. Since the potential energy, U, under the homogeneous deformation is a function of the variables, a_1 to a_6, U is expanded for an increment of deformation, Δa_k, as

$$U_{\{a_k+\Delta a_k\}} = U_{\{a_k\}} + \left(\frac{\partial U}{\partial a_k}\right)_{\{a_k\}} \Delta a_k + \frac{1}{2} B_{km} \Delta a_k \Delta a_m$$

$$B_{km} = \left(\frac{\partial^2 U}{\partial a_k \partial a_m}\right)_{\{a_k\}} , \tag{2.3}$$

neglecting terms of order Δa^3 and higher. Then, the change in the energy, ΔU, has the relation,

$$\Delta U - \left(\frac{\partial U}{\partial a_k}\right)_{\{a_k\}} \Delta a_k = \frac{1}{2} B_{km} \Delta a_k \Delta a_m \ . \tag{2.4}$$

The left hand side indicates the difference between the energy necessary to deform from a_k to $a_k + \Delta a_k$ and the imaginary work done by the applied load at the state $\{a_k\}$ during the incremental deformation. If the difference is positive, the extra energy must be supplied to deform, which means that the state $\{a_k\}$ is stable. On the other hand, the crystal becomes unstable when additional deformation no longer requires extra energy. Therefore, the stability of a crystal can be evaluated by the sign of the minor determinant of the Hessian, B_{km}. In tension with the transverse constraint, the instability condition of the lattice is evaluated by the sign of B_{33} because the variables, except a_3, are constant. Based on the EAM potential, the sign of B_{33} changes at the nominal strain, $(a_3 - a_0)/a_0$, of 0.304 [5]. It is noteworthy that the instability strain in the idealized deformation is much larger than the global strain at the cleavage fracture obtained by molecular dynamics simulation.

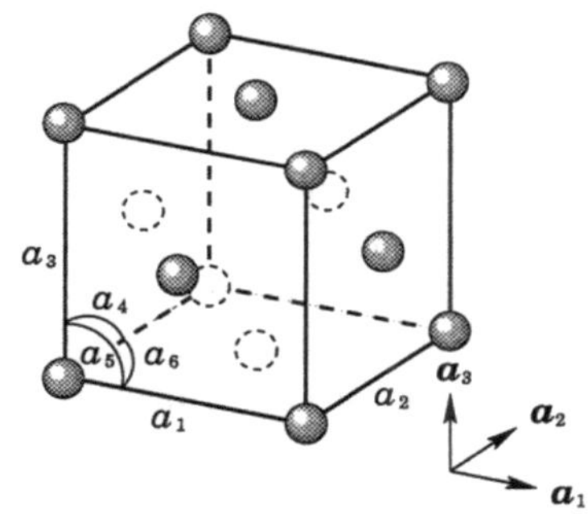

Fig. 2.4. Unit cell of fcc and variables, a_1 to a_6, which represent the uniform deformation of the crystal.

The critical strain in idealized deformation is indicated by the dotted line in Fig. 2.3. The cleavage crack nucleates at the instant when the maximum local strain between Planes 5 and 6 reaches the criterion. In general, the nucleation is dependent not only on $\varepsilon_{zz}^{(\mathrm{local})}$ but also another strain components and their derivations (*e.g.* $\partial\varepsilon_{xx}^{(\mathrm{local})}/\partial x$ and $\partial^2\varepsilon_{xx}^{(\mathrm{local})}/\partial x^2$) because each atom has its own freedom in the molecular dynamics simulation. However, $\varepsilon_{zz}^{(\mathrm{local})}$ gives the first approximation for the criterion of the nucleation of a cleavage crack, neglecting other factors.

The cleavage fracture along a twist grain boundary under the "strict constraint" was simulated by molecular dynamics in a previous paper [5], and the applicability of the criterion for the local strain was successfully verified.

3. Criterion on the Nucleation of a Dislocation

3.1 Shear Instability

3.1.1 Simulation Procedure. A simulation cell of a nickel single crystal illustrated in Fig. 3.1 is subjected to a shear strain, $\gamma_{zx}^{(\mathrm{global})}$, along the $[11\bar{2}]$ direction on the (111) planes, in which the atoms migrate due to glide of a partial dislocation. The molecular dynamics simulation is conducted by the repeat of the stepwise shear strain of 0.001 and the relaxation for 2.5 ps at 2.5 K [9]. The periodic boundary conditions are applied along the x- and y-directions, and the atoms on the top and bottom planes, indicated by the shaded area, are fixed on the lattice points and they are uniformly strained. The normal stresses, $\sigma_{xx}^{(\mathrm{global})}$, $\sigma_{yy}^{(\mathrm{global})}$ and $\sigma_{zz}^{(\mathrm{global})}$, are kept at zero during the simulation changing the shape and volume of the calculation cell.

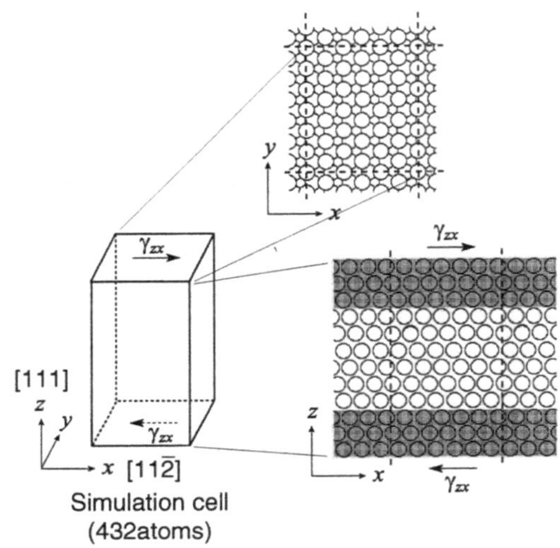

Fig. 3.1. Simulation cell in the calculation of shear instability, where [111] and $[11\bar{2}]$ are aligned along z- and x-axes, respectively.

3.1.2 Simulation Results and Local Instability Criterion. Figure 3.2 shows the relationship between the global shear stress, $\tau_{zx}^{(\mathrm{global})}$, and the global shear strain, $\gamma_{zx}^{(\mathrm{global})}$. After the elastic deformation, the stress drops sharply at $\gamma_{zx}^{(\mathrm{global})} = 0.176$. The snapshots of atomic arrangements before and after the stress drop are illustrated in Fig. 3.3. It shows that the atoms migrate toward the neighboring sites on the (111) plane producing a stacking fault. The migration takes place at once because there is no nucleation site of a dislocation. Figure 3.4 shows the distribution of local shear strain, $\gamma_{zx}^{(\mathrm{local})}$, evaluated at every lattice site [9]. The migration takes place at the plane with the maximum local strain.

In order to analyze the criterion, consider the ideal deformation of the crystal consisting of cells as illustrated in Fig. 3.5. The deformation here is described by the variables, a_1' to a_6', instead of a_1 to a_6 in the analysis of the previous section. When the magnitude of a_5' is changed while periodic

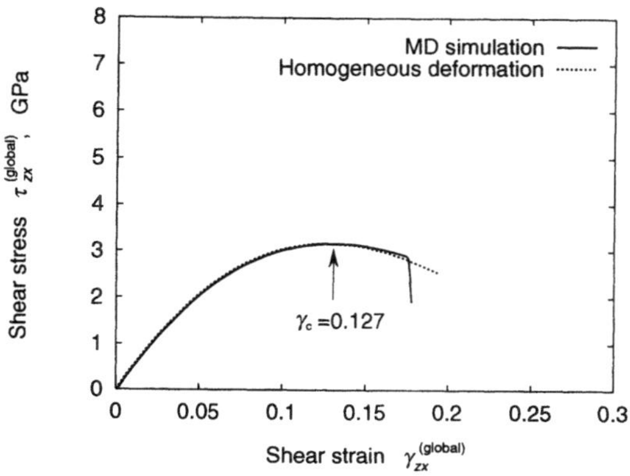

Fig. 3.2. Relationships between global stress and global strain under shear.

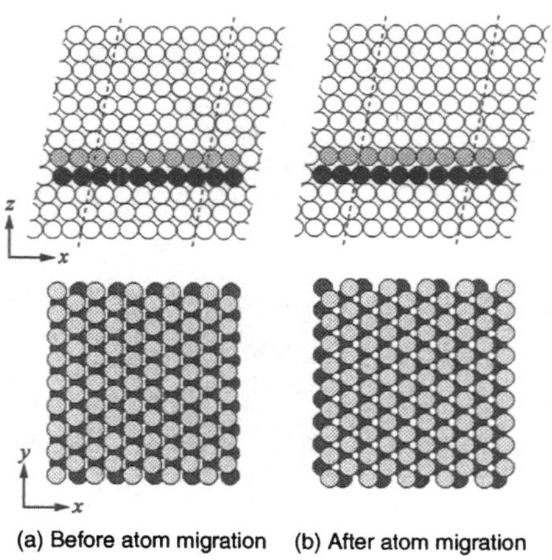

Fig. 3.3. Snapshots of atoms at yield.

boundary conditions applying, the crystal is subjected to a homogeneous shear strain. The normal stresses, $\sigma_{xx}^{(\mathrm{global})}$, $\sigma_{yy}^{(\mathrm{global})}$ and $\sigma_{zz}^{(\mathrm{global})}$, are kept at zero changing the variables, a_1' to a_3', by trial and error keeping $a_4' = a_6' = \pi/2$. The dotted line in Fig. 3.2 points out the $\tau_{zx}^{(\mathrm{global})}$–$\gamma_{zx}^{(\mathrm{global})}$ relation in the idealized homogeneous deformation. The relation coincides very well with the result in the MD simulation until the yield point. The instability condition in the idealized deformation of the crystal can be analyzed by a similar procedure to that described in the previous section on the basis of the Hessian, $B_{km}' = (\partial^2 U/\partial a_k' \partial a_m')$. The critical strain is found to be 0.127 in the idealized deformation, which coincides with the peak of the stress. This means that the crystal begins to deform without the extra stress. Then, the shear strain is localized by the instability in the molecular dynamics simulation until the slip takes place at $\gamma_{zx}^{\mathrm{global}} = 0.176$. This points out that the shear instability of the lattice is the trigger for the onset of slip.

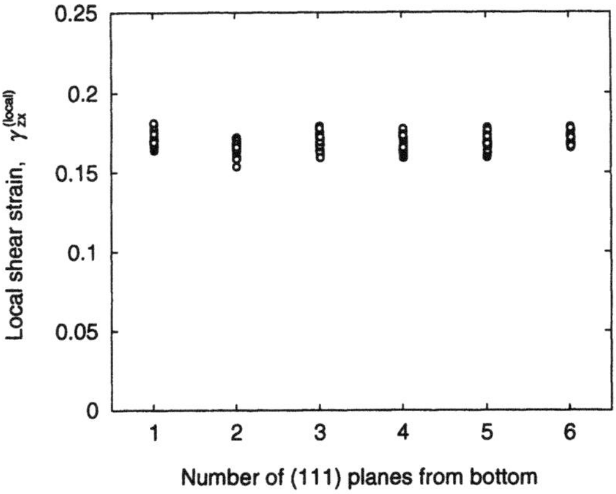

(a) $\gamma_{zx}^{(\mathrm{global})} = 0.175$

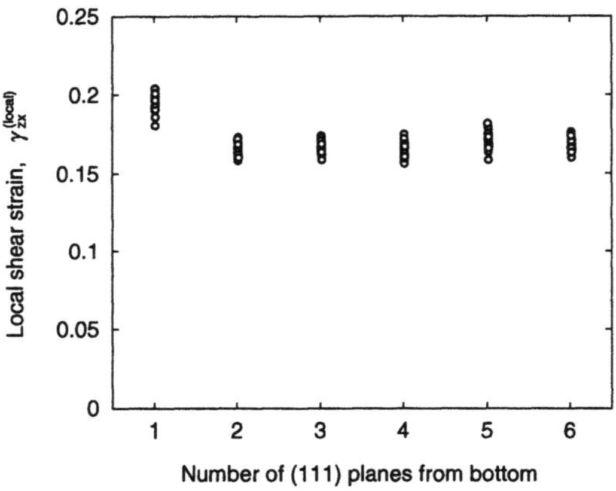

(b) $\gamma_{zx}^{(\mathrm{global})} = 0.176$

Fig. 3.4. Local shear strain on (111) plane. Abscissa represents number of (111) planes from the bottom of the simulation cell illustrated in Fig. 3.1.

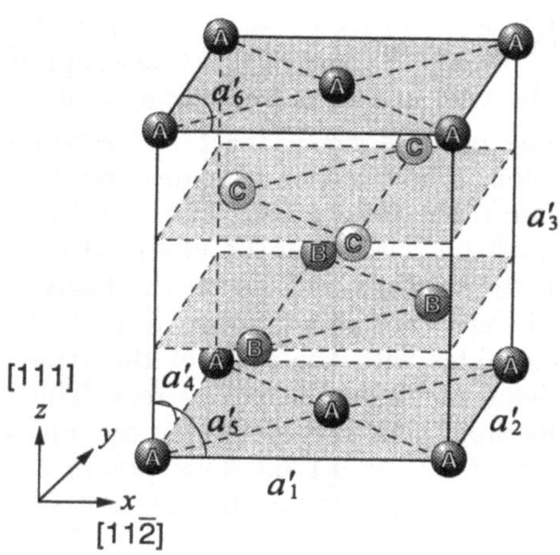

Fig. 3.5. Cell used for the instability analysis under idealized shear deformation.

3.2 Nucleation of a Dislocation

3.2.1 Simulation Procedure. Applying periodic boundary conditions on the simulation cell along the z-axis ([001] direction), as illustrated in Fig. 3.6, a thin wire is subjected to the tensile strain along the [001] by the straining/relaxation procedure described in the section 2.1. In this wire a dislocation is nucleated at the surface.

3.2.2 Simulation Results and Criterion for Dislocation Nucleation. The $\sigma_{zz}^{(\text{global})}$–$\varepsilon_{zz}^{(\text{global})}$ relation is plotted by the broken line in Fig. 2.2. The stress is much lower than that under the tension with the transverse constraint. Figure 3.7 shows the snapshots of atoms on the (111) planes where the stacking fault appears at the yield point [9]. Atoms on the upper plane migrate to the neighboring sites. In the process, a partial dislocation nucleates from a surface and glides through the slip plane to the $[11\bar{2}]$ direction [9]. The figure also designates the atoms by the shaded and solid circles at which the local shear strain reaches the instability criterion, $\gamma = 0.13$, evaluated in the previous section. The increase of the local shear strain, which is caused by the shear instability of the lattice, $\gamma_c = 0.127$, brings the high local strain area along the left surface and propagates it inward [9]. A partial dislocation nucleates and glides following the area. In other words, the nucleation of the dislocation is physically understood as the local shear instability along the surface and the nucleation criterion is approximately given by the condition that the local shear strain reaches the instability criterion of the idealized crystal deformation.

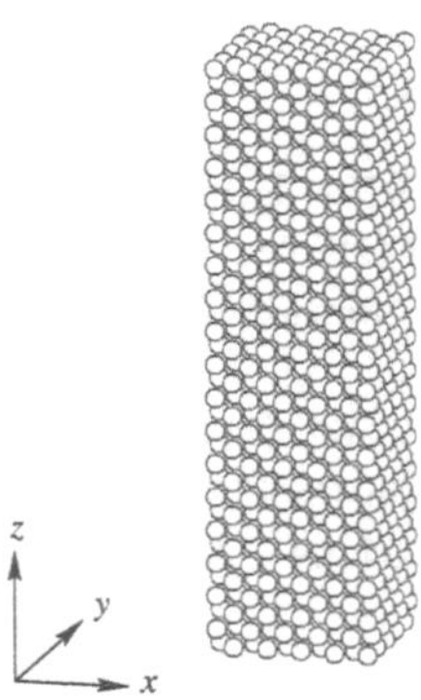

Fig. 3.6. Simulation cell in the calculation of dislocation nucleation.

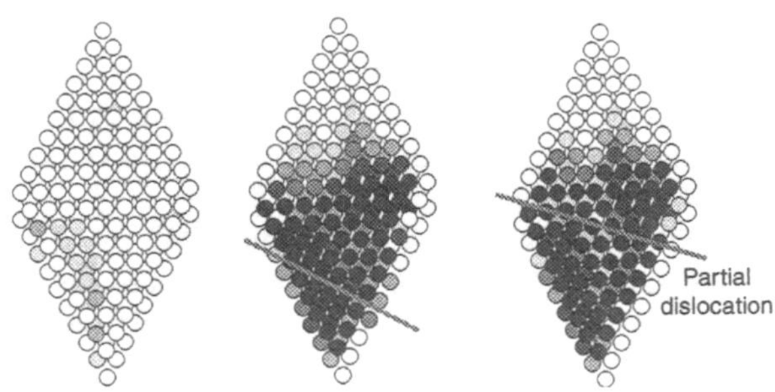

(a) $t = 0$fs (b) $t = 1750$fs (c) $t = 2000$fs

Fig. 3.7. Snapshots of atoms on the (111) plane at the yield point. Shaded circles indicate the atoms where the shear strain exceeds $\gamma = 0.13$.

4. Conclusion

The nucleation processes of a cleavage crack and a dislocation in a single crystal are simulated by molecular dynamics. The results obtained are summarized as follows;

(1) The brittle fracture takes place in tension of a small crystal with the strict constraint of transverse deformation. The cleavage crack nucleates at the lattice plane with the maximum tensile strain. The criterion for the nucleation is given by the local strain as a first approximation, which coincides well with the tensile strain at the instability in the idealized homogeneous deformation of the crystal.

(2) The slip instability is simulated by the molecular dynamics and a stacking fault is formed on a (111) plane at the maximum shear strain. Although the criterion evaluated from the local shear strain is much larger than the shear instability in the idealized homogeneous deformation of the crystal, the result suggests that the shear instability brings about localization of the shear strain to cause slip in actual materials.

(3) The ductile fracture is brought about by the glide of dislocations during straining of the wire. The partial dislocation nucleates on a (111) plane from a side surface along which the resolved shear strain is higher than that in other regions of the crystal. The critical shear strain coincides well with the shear instability strain obtained in (2).

References

1. For example, R.Chang: Int. J. Fract. Mech. **6**, 111–125 (1970). H.Kitagawa, A.Nakatani, Y.Shibutani: Mater. Sci. Eng. **A 176**, 263–269 (1994).
2. M.Hashimoto, Y.Ishida, R.Yamamoto, M.Doyama: Acta Metall. **32**, 1-11 (1984).
3. T.Kitamura, K.Yashiro, R.Ohtani: JSME Int. J., **40**–A, 430–435 (1997).
4. T.Kitamura, K.Yashiro, R.Ohtani: Proceedings of the Asia-Pacific Symposium on Advances in Engineering Plasticity and Its Applications - AEPA'96, Hiroshima, 1996, pp.541–546.
5. K.Yashiro, T.Kitamura, R.Ohtani: Trans. Jpn. Soc. Mech. Eng. **63**–A, 2380–2386 (1997) (in Japanese).
6. M.S.Daw, M.I.Baskes: Phys. Rev. B **29**, 6443–6453 (1984).
7. H.Kitagawa, A.Nakatani: Trans. Jpn. Soc. Mech. Eng. **59**–A, 256–262 (1993) (in Japanese).
8. F.Milstain: Phys. Rev. B **3**, 1130–1141 (1971).
9. T.Kitamura, K.Yashiro, R.Ohtani: (In preparation).

Part 4

Defects and Interfaces

Atomic Migration and Dynamic Rearrangement near Al Grain Boundary

Ken-ichi Saitoh[1] and Hiroshi Kitagawa[2]

[1]Department of Mechanophysics Engineering, Osaka University, Yamada-oka 2-1, Suita, Osaka, 565-0871 Japan
[2]Department of Adaptive Machine Systems, Osaka University, Yamada-oka 2-1, Suita, Osaka, 565-0871 Japan

Summary. A defect nucleated and grown by atomic migration in a crystalline structure influences the function of nanometer-scale wiring lines used in microelectronic devices. The first objective of this study is to investigate the basic mechanism of the failure due to the stress-induced migration by molecular dynamics simulation based on the effective-medium theory (EMT). The dependence of diffusion properties on atomic structures, temperature, and applied stress are investigated, and the largest diffusion coefficient is obtained near the junction between the grain boundary and the surface. The second objective is to elucidate the mechanism of atomic rearrangement regarded as slip on {111} planes near the junction. Under tensile loading perpendicular to the grain boundary, the concentration of slip brings about subsequent decohesion along the grain boundary. Moreover, a strong constraint due to the substrate and the passivation is taken into account in the simulation of slip.

Key Words. Molecular dynamics, stress-induced migration, effective-medium theory, grain boundary, surface, diffusion coefficient, scientific visualization

1. Introduction

Grain boundaries often dominate the strength of a polycrystalline metal. In particular, atomic-scale phenomena such as segregation, migration, and sliding affect the mechanism of defect nucleation at the grain boundary. Recent miniaturization of the components for electronic devices requires precise evaluation of the strength of micromaterials[1]. As aluminum lines in the devices, of which the width is only several hundred nanometer or smaller, are commonly used, enormous electron current densities cause unexpected failure due to electromigration (EM). The stress induced by thermal expansion mismatch among the lines, silicon substrate, and passivation film (Fig. 1.1a) produces migration as well (stress-induced migration, SM) (Fig. 1.1b). Since EM and SM lead to breakage or functional deterioration of the lines, many studies have been carried out on this aspect of reliability in the electronic devices [2–5].

Diffusion along a grain boundary is assumed to be dominant in failure by the EM and SM. The bamboo-like grain structure is used for preventing the

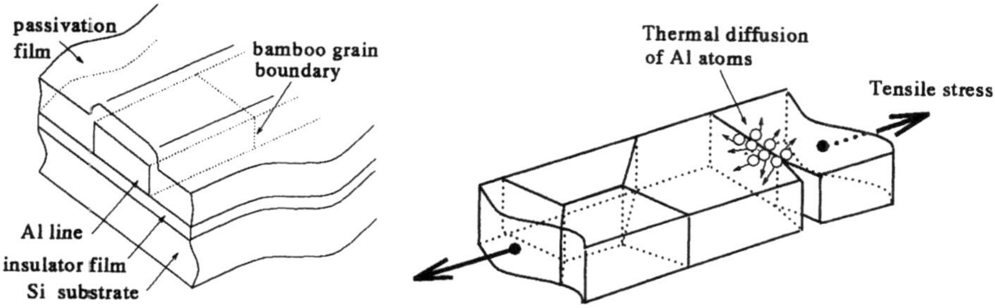

(a) Structure of aluminum line (b) SM mechanism near grain boundary

Fig. 1.1. Aluminum line in electronic device and its failure mechanism by SM.

diffusion due to the current along the line and a defect is formed due to the SM at the junction between the grain boundary and the passivation. In order to elucidate the SM process, some atomistic studies at the junction region have been carried out [6–9].

The objective of the present study is to investigate the atomic migration mechanism by atomic-scale computer simulation. The effective-medium theory (EMT)[10] is used for the evaluation of the interatomic force in the molecular dynamics analysis, because it is capable not only of precisely estimating the grain boundary energy and the surface energy but also of reproducing an approximate result of diffusion [8, 9] and dislocation motion[11, 12].

In this study, we also investigated dynamic rearrangement of the atomic structure near the junction under tensile loading. We found that slip near the junction is also important as another mechanism of failure, although it does not incorporate the thermal activation process.

2. Molecular Dynamics Simulation Based on Effective-Medium Theory

Molecular dynamics (MD) simulations are conducted based on the effective-medium theory (EMT). Since incorporation of the many-body effect due to variation of electron density is required to obtain accurate results in an investigation of an inhomogeneous atomic structure, such as a grain boundary or a surface, the EMT formulated by Jacobsen[10] is adopted here. The outline is briefly shown as follows.

The total energy of atomic system, E_{tot}, is approximately given by

$$E_{\text{tot}} = \sum_{i=1}^{N} \left[E_{c,i}(\bar{n}_i) + \Delta E^{\text{corr}}_{\text{AS},i} \right] . \tag{2.1}$$

Here, $E_{c,i}(\bar{n}_i)$ is the change in energy when the i-th atom is embedded in a homogeneous electron gas (effective medium) produced by the surrounding atoms. The second term corrects the approximation overlap of the electron density in the neighboring atoms. $E_{c,i}$ and an average electron density, $\bar{n_i}$, are calculated by

$$E_{c,i}(\bar{n_i}) = E_0 + E_2\left(\frac{\bar{n_i}}{n_0} - 1\right)^2 + E_3\left(\frac{\bar{n_i}}{n_0} - 1\right)^3, \tag{2.2}$$

$$\bar{n_i} = n_0 \exp(\eta s_0)\left(\frac{\sigma_{1,i}}{12\gamma_1}\right)^{\frac{\eta}{\beta\eta_2}}, \tag{2.3}$$

$$\sigma_{1,i} = \sum_{j\neq i} \exp(-\eta_2 r_{ij})\theta(r_{ij}). \tag{2.4}$$

The correction term, $\Delta E^{\mathrm{corr}}_{\mathrm{AS},i}$, is evaluated by

$$\Delta E^{\mathrm{corr}}_{\mathrm{AS},i} = \alpha(\bar{n_i} - \bar{n_i}^{\mathrm{fcc}}), \tag{2.5}$$

$$\bar{n_i}^{\mathrm{fcc}} = n_0 \exp(\eta s_0)\frac{\sigma_{2,i}}{12\gamma_2}, \tag{2.6}$$

$$\sigma_{2,i} = \sum_{j\neq i} \exp\left(-\frac{\eta}{\beta} r_{ij}\right)\theta(r_{ij}), \tag{2.7}$$

where $\bar{n_i}^{\mathrm{fcc}}$ is an average electron density of a perfect crystal with fcc structure. $\theta()$ is a step function to connect each electron density smoothly. Parameters required for aluminum are shown in [10] and this relies only on non-empirical parameters.

The details of atomic models and boundary conditions of the simulation in this study are described in the following sections.

3. Results and Discussions

3.1 Diffusion Properties of Aluminum along a $\Sigma = 5(100)$ Symmetrical Tilt Grain Boundary

It is recognized that atomic diffusion processes play a dominant role in SM. A void or a groove is initiated near the junction where a grain boundary and a surface (or interface between the lines and passivation film) meet (Fig. 1.1b). Diffusion coefficients (DCs) are estimated for the computer model illustrated in Fig. 3.1[9].

Using the coincidence site lattice (CSL) theory[13, 14], which is based on the periodicity of the grain boundary, the grain boundary simulated here is represented as a $\Sigma = 5(100)$ symmetrical tilt grain boundary (STGB). A $\Sigma = 9(110)$ STGB and a $\Sigma = 5(100)$ asymmetrical tilt grain boundary (ATGB) are also studied in the next section.

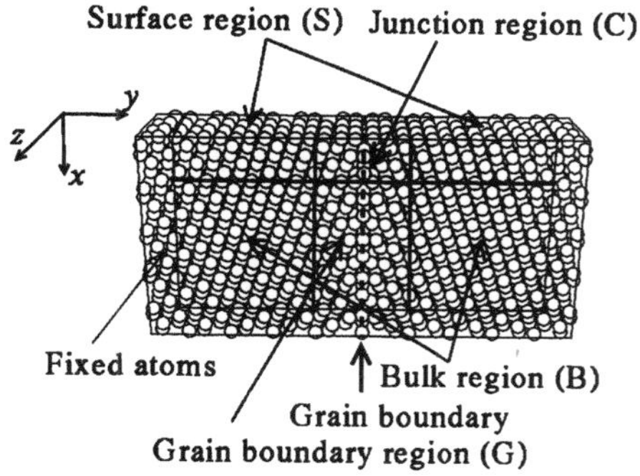

Fig. 3.1. Atomic model for investigation of diffusion properties.

In the model, a $\Sigma = 5(100)$ STGB and the free surface are located on the planes parallel to xz and yz planes, respectively. The forth-order Gear algorithm is used for numerical integration of the equation of motion. The MD time step is 2.0 fs.

DCs are estimated from the mean-square displacement (MSD) of atoms $\langle r^2(t)\rangle$ using the relation,

$$D = \frac{1}{6t}\langle r^2(t)\rangle. \tag{3.1}$$

The displacement, $r(t)$, is obtained for 0.40 ns by the MD simulation. We divide the structure into four regions, C (junction), S (surface), G (grain boundary), and B (bulk) as shown in Fig. 3.1. The temperature dependencies of the DCs, which are the average in each region, are shown in Fig. 3.2. The horizontal axis in Fig. 3.2 is the reciprocal of temperature normalized by the melting temperature, T_m^{calc}, evaluated by another MD calculation. The C region has the largest value at any temperature, which means marked atomic motions occur near the junction. Activation energies for self-diffusion, Q, are estimated, using the following Arrhenius relation,

$$D = D_0 \exp\left(-\frac{Q}{k_B T}\right), \tag{3.2}$$

where D_0 is a material constant and k_B is the Boltzmann constant. The activation energies obtained are listed in Table 3.1. This indicates that the energy in the C region is about one-third of that in the B region.

Table 3.1. Regional activation energy for diffusion

Region	C	S	G	B
Activation energy Q (eV)	0.341	0.579	0.578	1.008

In the actual aluminum lines, SM is driven by high tensile stress induced by thermal mismatch. The stress reaches about several hundred MPa. Then, the dependence of DCs on the strain along the line is examined by the MD simulations. Fig. 3.3 shows the relationship between the DC and applied

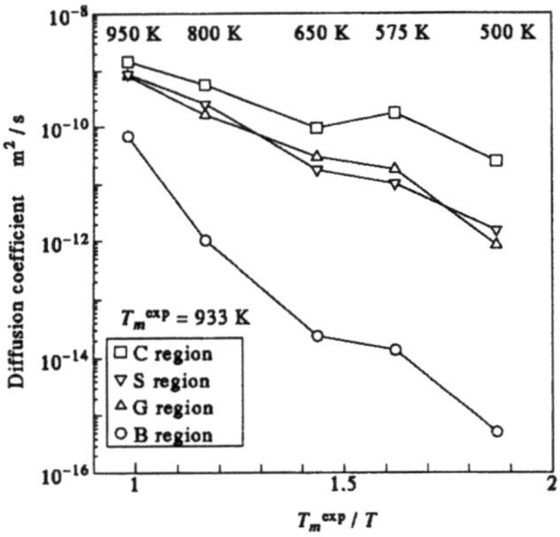

Fig. 3.2. Regional diffusion coefficient.

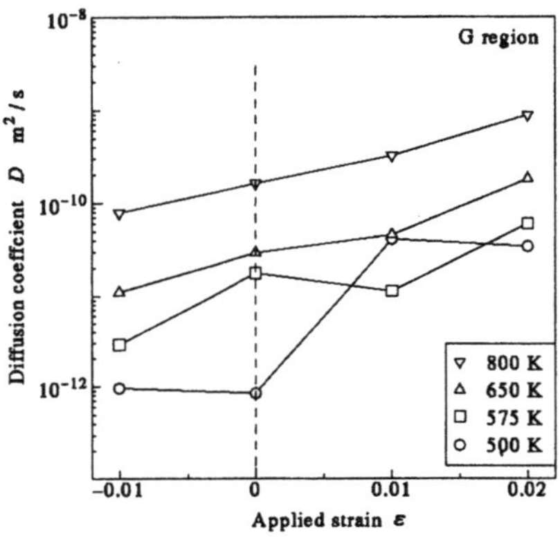

Fig. 3.3. Relationship between strain and diffusion coefficient in the G region.

strain in the G region. DC increases exponentially with tensile strain and decreases with compressive strain.

3.2 Atomic Dynamic Rearrangement near Aluminum Grain Boundary

Free surfaces, grain boundaries, and their junctions are supposed to have high atomic mobility. Therefore, deformation due to slip is accelerated in such regions and this may lead to the failure of the lines. During catastrophic decohesion along the grain boundary, rapid rearrangement of atomic structure takes place[15]. The process of dynamic rearrangement is investigated in the following sections.

3.2.1 Effect of Grain Boundary Structure on Atomic Rearrangement. It is well-known that a grain boundary acts as a sink and source of dislocations. Therefore, the deformation due to the dislocation slip is influenced by the grain boundary structure as well as the orientation of grains. Three kinds of grain boundary, a $\Sigma = 5(100)$ STGB, a $\Sigma = 9(110)$ STGB, and a $\Sigma = 5(100)$ ATGB are subjected to tensile strain perpendicular to the grain boundary plane as illustrated in Fig. 3.4. In the model, periodic boundary conditions are applied in the y and z directions and the ends of the grain boundaries meet the surface. After stepwise straining of 2%, it is relaxed for 2 ps. The procedure is repeated until 18% total strain is reached. The temperature is controlled to within 5.0 K by the velocity scaling method. Fig. 3.5 shows the relation between strain and normal (σ_x and σ_y) and shear (τ_{xy}) stresses on the cell. In the two $\Sigma = 5$ grain boundary models, σ_y suddenly decreases at over 10% of strain whereas the decrease is not observed in the $\Sigma = 9$ grain boundary model.

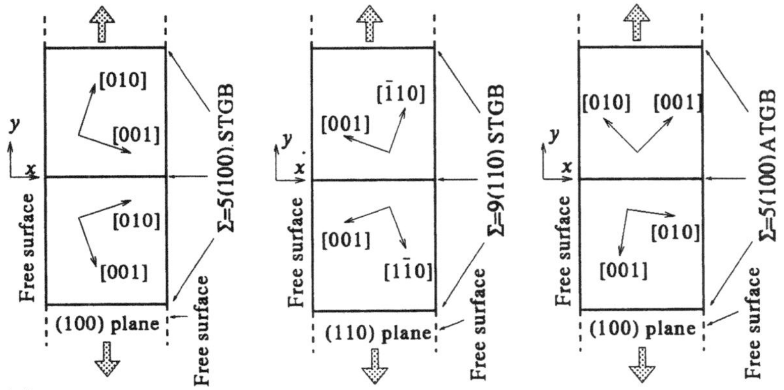

(a) $\Sigma = 5(100)$ STGB (b) $\Sigma = 9(110)$ STGB (c) $\Sigma = 5(100)$ ATGB

Fig. 3.4. Three atomic models of grain boundary and free surface.

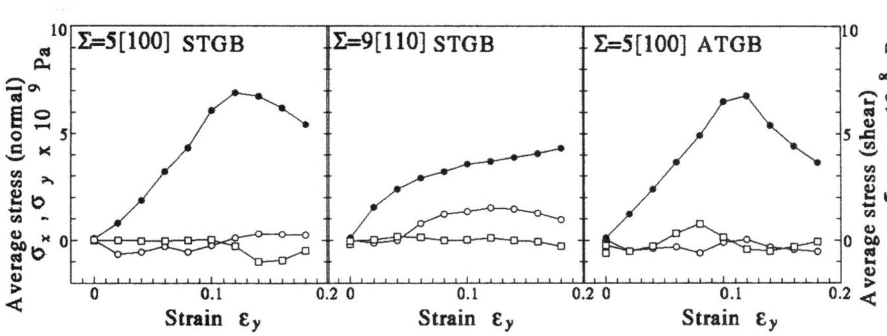

(a) $\Sigma = 5(100)$ STGB (b) $\Sigma = 9(110)$ STGB (c) $\Sigma = 5(100)$ ATGB

Fig. 3.5. Transition of average atomic stresses.

Rearrangement of the atomic structure is mainly caused by slip on $\{111\}$ planes in fcc structure. For quantitative estimation of the slip, the relative slip vector is defined as

$$\boldsymbol{u}_{ij} = (\boldsymbol{r}_i(t) - \boldsymbol{r}_j(t)) - (\boldsymbol{r}_i(t_0) - \boldsymbol{r}_j(t_0)), \tag{3.3}$$

where $\boldsymbol{r}_i$ and $\boldsymbol{r}_j$ are the position vector of i-th and j-th atoms, respectively. $\boldsymbol{r}_i(t_0)$ and $\boldsymbol{r}_j(t_0)$ are those in a reference configuration, which is defined as the atomic structure at no strain. Thus, the relative displacement between atomic layers can be visualized as in Fig. 3.6. In Fig. 3.6, relative slip up to strain 6% and 14% is depicted using lines corresponding to the direction and magnitude of relative slip vectors. Thick lines in the figure indicate the concentrated area of slip. In a $\Sigma = 5(100)$ STGB, slip concentrates in the junction region and decohesion along the grain boundary takes place, whereas slip spreads into the perfect crystal region of grains and no decohesion occurs in the $\Sigma = 9(110)$ STGB. In the $\Sigma = 5(100)$ ATGB the difference of contraction above and below the grain boundary plane promotes high shear stress along the grain boundary and leads to decohesion of the atomic system. Two layers of (100) or (110) planes in the $\Sigma = 5$ or $\Sigma = 9$ grain boundary models are shown in Fig. 3.7b,c. The Burgers' vector of the latter is smaller than that of the former. It is concluded that the direction of the rotation axis of the grain boundary (in the z direction) is one of the important factors which control slip length and direction.

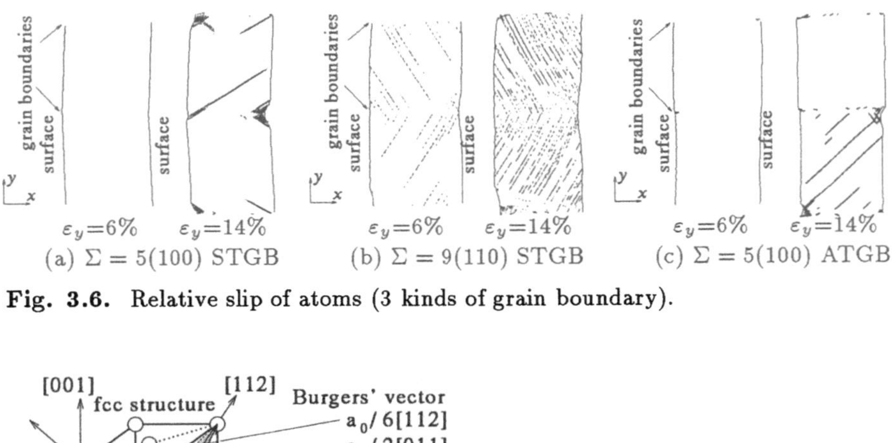

(a) $\Sigma = 5(100)$ STGB (b) $\Sigma = 9(110)$ STGB (c) $\Sigma = 5(100)$ ATGB

Fig. 3.6. Relative slip of atoms (3 kinds of grain boundary).

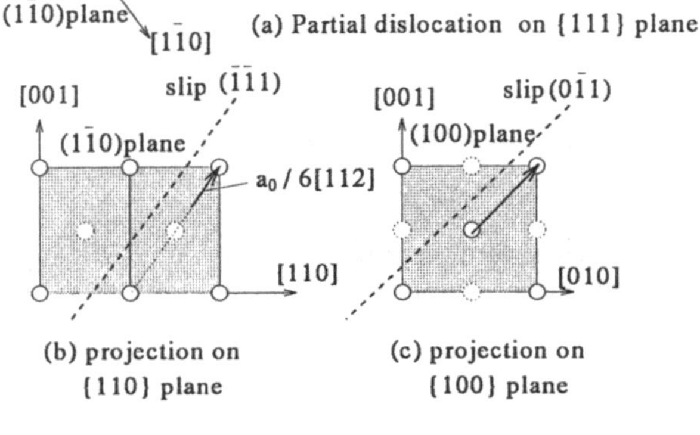

Fig. 3.7. Explanation of slip plane and direction of the models ((b) for $\Sigma = 9$ grain boundary and (c) for $\Sigma = 5$ grain boundary).

3.2.2 Effect of Surface Constraint on Atomic Rearrangement. An aluminum line inside an electronic device is surrounded by a passivation film (Fig. 1.1) and is exposed to a high temperature environment (200-300°C). Therefore, it is supposed that the failure is caused by brittle cracking due to high stress promoted by the constraint of deformation[2]. In order to investigate the effect of the constraint on the rearrangement of the atomic structure near the grain boundary, the MD simulation are conducted on the atomic structures with the junction as shown in Fig. 3.8. They are subjected to tensile strain under different constraint conditions on the surface (left side edge along the y direction in Fig. 3.8). In model A (Fig. 3.8a), the surface is covered by atoms with fixed displacement. In this case, crystalline orientation is continuous in the interface between the fixed region and the free region inside. In model B (Fig. 3.8b), another grain is set at the fixed region. In model C (Fig. 3.8c), the grain boundary is connected to the free surface. Since a groove at the junction is hard to nucleate, except in model C, a starter void is introduced in advance of the junction, as shown in Fig. 3.8a,b.

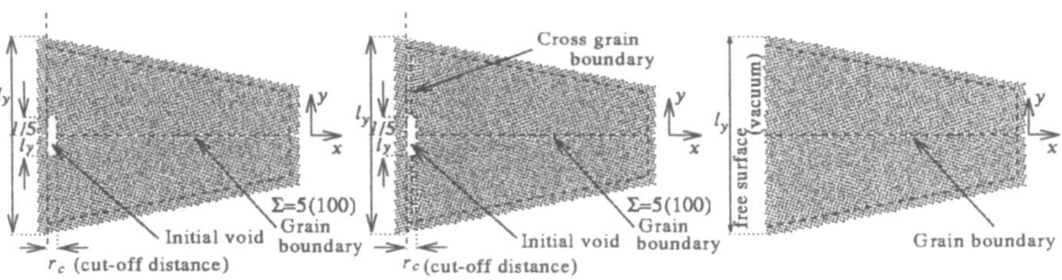

(a) model A (b) model B (c) model C

Fig. 3.8. Atomic models of junction structure.

The velocities of atoms in the fixed region outside the dashed lines in Fig. 3.8 are prescribed by the applied tensile or compressive strain. The strain rate in the y direction, $\dot{\varepsilon}_y$, is $4.0 \times 10^9\,\mathrm{s}^{-1}$ and that in the x direction is given by $\dot{\varepsilon}_x = -(C_{12}/C_{11})\dot{\varepsilon}_y$, where C_{11} and C_{12} are the elastic constants taking into account the crystalline anisotropy. Although these strain rates are enormously high, the maximum velocity of atoms, which is about $20\,\mathrm{ms}^{-1}$, is reasonable for observation of dynamic change in atomic structure [11]. Temperature is controlled to within 450 K using the velocity scaling method.

Figure 3.9 shows the atomic configuration near the junction region at ε_y of 12%. In the figures, the atoms are shown by circles and the atoms with high relative slip displacement are presented by shading. Here, the displacement is evaluated as the summation of $|\boldsymbol{u}_{ij}|$, which is introduced by (3.3), with respect to the nearest-neighbor atoms, and is given by

$$s_i = \sum_{j}^{neighbor} |\boldsymbol{u}_{ij}|. \tag{3.4}$$

The darkness of shading in Fig. 3.9 represents the levels of displacement at b, $2b$, and $3b$. The darkest one indicates the largest relative slip displacement.

In model A, the amount of slip induces grain boundary migration. In model B, slip concentrates both on two crossing grain boundaries and decohesion occurs along the grain boundary perpendicular to the tensile direction. Model C presents the same tendency as model B.

Dynamic structural change in these models has been monitored and observed as an animation (movie-file in computer or video) which consists of a snapshot of the coordinate data every 400 fs. These videos are helpful to visualize the dynamic structural change in detail.

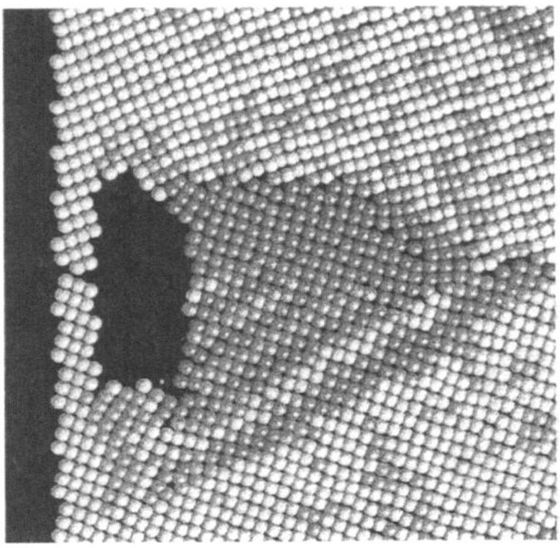

(a) model A

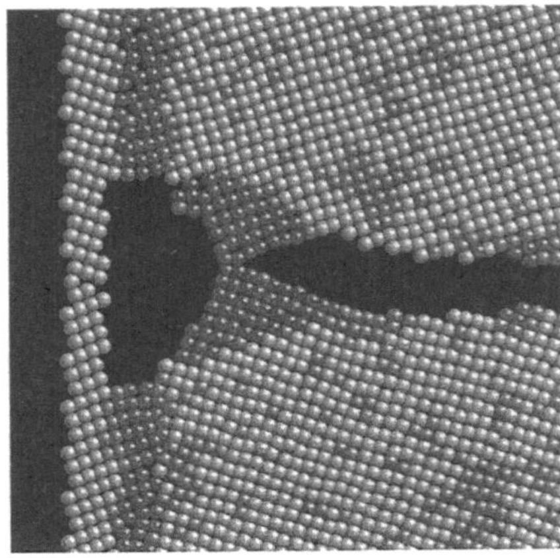

(b) model B

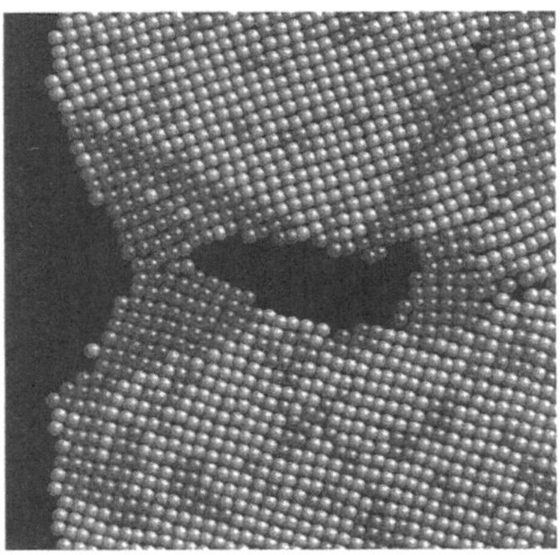

(c) model C

Fig. 3.9. Visualization of atomic configuration and relative slip (ε_y is 12%).

Fig. 3.10 shows the frequency of relative slip in the four regions, C, S, G, and B shown in Fig. 3.1. As displayed typically in Fig. 3.10a, there are two peaks in the relative slip. A peak near zero magnitude corresponds to atomic movement which does not accomplish slip motion. On the other hand, when slip on a {111} plane brings about $s_i = 3b$, where b is the Burgers' vector

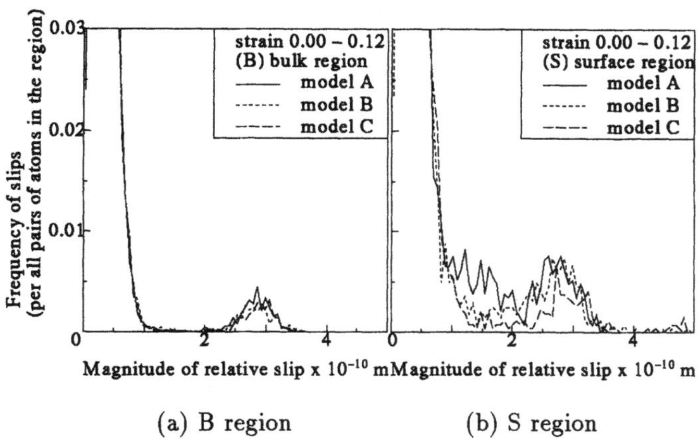

Fig. 3.10. Relationship between magnitude of relative slip and their frequency (relative slip is estimated from $\varepsilon_y = 0.0 \sim 0.12$).

for a perfect edge dislocation, another peak near $b = \sqrt{2}/2a_0$ appears in the relative slip. Here, a_0 is the lattice constant and $\sqrt{2}/2a_0$ equals 2.9×10^{-10} m in aluminum.

The second peak represents a stable configuration of atoms after the slip of a perfect dislocation. However, in region S, which contains the interface or the free surface, peaks of model A are not so distinctive and slip with intermediate displacement dominates the rearrangement. This is because the magnitude of the Burgers' vector of an imperfect dislocation (partial dislocation) exists between the two peaks and such imperfect slip is accumulated near the interface regions due to the strong constraint condition in the case of model A. This leads to a suppression of atomic rearrangement near the grain boundary and prevents a subsequent decohesion.

4. Conclusion

Diffusion properties and dynamic rearrangement near an aluminum grain boundary are investigated by molecular dynamics based on the effective-medium theory (EMT). The results obtained in this study are as follows.

- Diffusion coefficient in the junction is the largest and its activation energy is about one-third of that in the bulk region.
- Slip is one of the characteristic atomic migration mechanisms in the fcc structure. The free surface and grain boundaries accelerate it. In $\Sigma = 5(100)$ STGB and ATGB, slip concentrated in the junction or the grain boundary. This brings about decohesion along the grain boundary, whereas slip spreads into the perfect crystal region and no decohesion takes place in a $\Sigma = 9(110)$ STGB.

- When a strong constraint of deformation exists near the junction, the rearrangement of atoms is suppressed and the ratio of slip with intermediate Burgers' vectors increases.

References

1. T.Kitamura, K.Yashiro, R.Ohtani: Proceedings of the 3rd Asia-Pacific symposium on advances in engineering plasticity and its applications (eds. T. Abe and T. Tsuta), pp.541-546, (Pergamon, 1996).
2. K.Hinode, N.Owada, T.Nishida, K.Mukai: J. Vac. Sci. Technol. B **5**, 518-522 (1987).
3. A.S.Oates, D.L.Barr: J. Electron. Mater. **23**, 63-66 (1994).
4. T.Onishi: Materia Japan (in Japanese) **33**, 755-759, (1994).
5. H.Kaneko, T.Kawanoue: Materia Japan (in Japanese) **36**, 306-310, (1997).
6. T.Iwasaki, N.Sakaki, N.Chiba, Y.Abe: Trans. Jpn. Soc. Mech. Eng. A (in Japanese) **61**, 1376-1382 (1995).
7. T.Kitamura, R.Ohtani, T.Yamanaka, K.Yashiro, JSME Int. J., Ser.A **39**, 291-296 (1996).
8. H.Kitagawa, A.Nakatani, S.Ogata, K.Saitoh, Y.Maegawa: Trans. Jpn. Soc. Mech. Eng. A (in Japanese) **62**, 853-858 (1996).
9. H.Kitagawa, A.Nakatani, S.Ogata, K.Saitoh,, Y.Maegawa, JSME Int. J., Ser.A **40**, 203-210 (1997).
10. K.W.Jacobsen, J.K.Nørskov, M.J.Puska: Phys. Rev., B **35**, 7423-7442 (1987).
11. J.Schiøtz, K.W.Jacobsen, O.H.Nielsen: Philos. Mag. Lett. **72**, 245-250 (1995).
12. J.Schiøtz, T.Rasmussen, K.W.Jacobsen, O.H.Nielsen: Philos. Mag. Lett. **74**, 339-344 (1996).
13. S.Ranganathan: Acta Crystallogr. **21**, 197-199 (1966).
14. G.A.Chadwick, D.A.Smith (eds.): *Grain boundary structure and properties*: Ch.5, pp.139-200, (Academic Press, 1976).
15. H.Kitagawa, A.Nakatani, S.Ogata, K.Saitoh: Proceedings of the 3rd Asia-Pacific symposium on advances in engineering plasticity and its applications (eds. T. Abe and T. Tsuta), pp.535-540, (Pergamon, 1996).

Discrete Dislocation Dynamics Study on Distribution of Dislocations and Stress Field near Crack Tip

Akihiro Nakatani[1], Hiroshi Kitagawa[1], and Makoto Sugizaki[2]

[1]Department of Adaptive Machine Systems, Graduate School of Engineering, Osaka University, 2-1, Yamadaoka, Suita, Osaka 565-0871, Japan
[2]Department of Mechanical Engineering, Graduate School of Engineering, Osaka University, 2-1, Yamadaoka, Suita, Osaka 565-0871, Japan

Summary. The discrete dislocation dynamics (DDD) method is a suitable computer simulation procedure which analyzes problems at the mesoscopic scale. In the method, many dislocations are driven dynamically by the forces which result from the interaction of dislocations. This method can not only deal with the movement of many dislocations in a relatively large area during relatively long time periods, but also certain interesting information may be obtained, when the elementary processes of dislocations are explicitly studied by some atomic simulations, ex. molecular dynamics. In this paper, the DDD simulations are carried out to obtain the mechanical and structural state near the crack tip region. The model with a crack is assumed to be bcc iron. The crack front is taken in the $[1\bar{1}0]$ direction. Two situations are considered, in the first the crack surface on a (110) plane and in the second a (001) plane is selected. On the assumption of plane strain, the number of active slip systems is confined to two of the 48 proper bcc slip systems; one has a $(\bar{1}\bar{1}2)$ slip plane and a [111] slip direction, and another has a (112) slip plane and $[11\bar{1}]$ slip direction. At the begining of the simulation, the edge dislocations which are related to these slip systems, are scattered randomly. The relaxation calculation is performed to obtain an equilibrium state of the system. Then, a stable state is determined under mode I loading to obtain an equilibrium of the specified boundary conditions. The shield and amplification of stress intensity factor are observed in the obtained stress distribution. The distribution of plastic strain, which is roughly evaluated by the path of dislocation movement, qualitatively corresponds to the solution based on macroscopic crystal plasticity theory.

Key Words. Discrete dislocation dynamics, dislocation, mode I loading, plastic deformation, shielding of stress intensity factor

1. Introduction

Understanding how fracture occurs and what are the quantitative parameters (driving force) which specify the fracture process is very important for estimating the strength of materials. The structural integrity is evaluated by comparing the practically expected driving force with proper critical values of the material strength (resistance against fracture). Most of conventional structural design has been done by estimating, first, the driving force by continuum mechanics as precisely as possible, and then comparing it with the resistance obtained by experiments. Recently, the needs of design of microscale structures in which the continuum approach is not suitable have been recognized as a result of the artificial product which consists of elements with sub-micrometer or nanometer scale.

In the microscopic region, molecular dynamics simulation can precisely predict the structure and character of materials [1]. This implies that resistance under an ideal condition can be obtained. Therefore, if the practical driving force is estimated precisely, we will be able to carry out a design by computer alone. However, this is not easy. It is difficult to analyze all of the objects and the circumstances which surround it from the atomic level. Any simulation has to be done by handling some partial system. A good estimation of the driving force is achieved under good modeling of the partial system. How to model the partial system is deeply concerned with the problem of how to specify the boundary conditions of the microscopic problem.

It is well known that even a cleavage crack which causes brittle fracture, involves a small plastic deformation at the crack tip [2]. Dislocation movement plays an important role in the fracture mechanism. The style of macroscopic fracture is determined by the possibility of propagation of the plastic and damage zones which are characterized by a dislocation, deformation twin, and a micro crack formed near the crack tip. These defect structures behave as filters, the crack tip field is isolated mechanically from the outside region (so-called shielding, or amplifying, of stress intensity factor). Because of these defects, even if the resistance to change in the atomic arrangement is precisely evaluated by atomic scale simulation, the corresponding driving force is not estimated well. As a result, the quantitative estimation of resistance becomes difficult formally.

For such problems, an intermediate scale between microscopic and macroscopic scales must be treated properly. In this paper, we discuss the simulation based on discrete dislocation dynamics [3, 4] (DDD), which is one of the mesomechanics procedures that deals with the intermediate region between macroscopic slip deformation and microscopic atomic arrangement in crystal structures. The method is applied to the region near the crack tip. The geometrical investigations are discussed by observing the movement and distribution of dislocations. The examination of shielding and amplifying effect of the stress intensity factor is accomplished by investigating the stress field.

2. Discrete Dislocation Dynamics

Atomic calculations [5, 6] indicated that the distribution of stress and displacement around a crack tip or dislocation agrees well with the solutions of linear elasticity except in the immediate vicinity. This seems to be one of the examples which the St.Venant Principle implies and this fact can establish the validity of approximation that treats a dislocation as a singular line. In a two-dimensional treatment, it becomes a singular point.

The DDD method is a direct simulation of dislocation movement under specific boundary conditions, in which the following stages are iterated; the force on each dislocation, which is induced by the interaction of dislocations, and the external force (body force and surface traction) are evaluated, and the dislocations are moved by an equation of motion.

2.1 Evaluation of Stress Field

A boundary value problem to determine the stress and displacement field in the finite body with dislocations is considered, in which the traction force and displacement are specified on S_t and S_u, respectively, as shown in Fig. 2.1a. The original problem is decomposed into two different problems A and B: Problem A is for an infinite body with dislocations as shown in Fig. 2.1b, and problem B is for the finite body without a dislocation as shown in Fig. 2.1c.

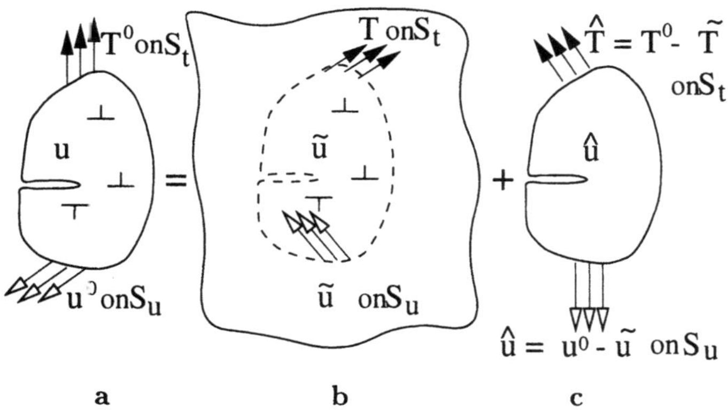

Fig. 2.1. Decomposition of Problem. (**a**) Original problem, (**b**) problem A, (**c**) problem B.

2.2 Problem A

If we assume that N dislocations are included in an infinite body, the stress field is obtained by superposition of stress which is induced by each dislocation as follows;

$$\tilde{\sigma}_{ij} = \sum_{\alpha=1}^{N} \sigma_{ij}^{\alpha}, \tag{2.1}$$

where, σ_{ij}^{α} is the stress induced by the dislocation α. In the same way, the displacement field is obtained by superposition as,

$$\tilde{u}_i = \sum_{\alpha=1}^{N} u_i^\alpha. \tag{2.2}$$

These equations give the solution of problem A. The closed form expressions for the stress and displacement which are induced by a dislocation in an infinite body are available [7].

2.3 Problem B

The surface traction on S_t is calculated by the solution of problem A with the help of Cauchy's stress formula, and $\tilde{\boldsymbol{u}}$ on S_u is given as the solution of problem A. The problem B is the boundary value problem without a dislocation under the complementary boundary conditions $\hat{\boldsymbol{T}}$ on S_t, and $\hat{\boldsymbol{u}}$ on S_u which give rise to the original boundary values after addition to those of problem A. The governing equations of the problem are expressed by,

$$\hat{\sigma}_{ij,j} = 0 \quad \text{in} \quad V \tag{2.3}$$

$$\hat{\epsilon}_{ij} = \frac{1}{2}\left(\hat{u}_{i,j} + \hat{u}_{j,i}\right) \quad \text{in} \quad V \tag{2.4}$$

$$\hat{\sigma}_{ij} = E_{ijkl}\hat{\epsilon}_{kl} \quad \text{in} \quad V \tag{2.5}$$

$$t_i = \hat{T}_i = t_i^0 - \tilde{T}_i \quad \text{on} \quad S_f \tag{2.6}$$

$$u_i = \hat{U}_i = u_i^0 - \tilde{U}_i \quad \text{on} \quad S_u, \tag{2.7}$$

where $\boldsymbol{E}$ is the elastic constant. This problem is usually solved by FEM.

2.4 Original Problem

The solution of the original problem is expressed by the superposition of problem A and B, given by,

$$\sigma_{ij} = \tilde{\sigma}_{ij} + \hat{\sigma}_{ij} \quad \text{in} \quad V \tag{2.8}$$

$$\epsilon_{ij} = \tilde{\epsilon}_{ij} + \hat{\epsilon}_{ij} \quad \text{in} \quad V \tag{2.9}$$

$$u_i = \tilde{u}_i + \hat{u}_i \quad \text{in} \quad V. \tag{2.10}$$

2.5 Evolution of Dislocation Movement

The dislocation is forced to move by an applied force to dislocation $\boldsymbol{f}^\alpha$ and resolved shear stress τ^α. The static equilibrium condition in which $\boldsymbol{f}^\alpha$ is vanishing is equivalent to the stationary condition of potential energy Π in the deformation field. The Peach–Koehloer force acting on the dislocation α

is given by,

$$f_i^\alpha = e_{ijk} t_j^\alpha \left\{ \left(\hat{\sigma}_{kl} + \sum_{\substack{\beta=1 \\ \beta \neq \alpha}}^{N} \sigma_{kl}^\beta \right) b_l^\alpha \right\}, \tag{2.11}$$

where $\boldsymbol{t}^\alpha$ is the unit direction vector along the dislocation line, $\boldsymbol{b}^\alpha$ is the Burgers' vector, e_{ijk} is the permutation symbol, and N is the total number of dislocations.

In the case of the edge dislocation in the xy plane, the component of the force along the slip direction $f^{\alpha(n)}$ is given by,

$$f^{\alpha(n)} = m_i^\alpha \left(\hat{\sigma}_{ij} + \sum_{\substack{\beta=1 \\ \beta \neq \alpha}}^{n} \sigma_{ij}^\beta \right) b_j^\alpha, \tag{2.12}$$

where $\boldsymbol{m}^\alpha$ is the normal vector of the slip plane of dislocation α. The force which drives the dislocation is expressed by,

$$f^\alpha = \tau^\alpha b^\alpha, \tag{2.13}$$

where $b^\alpha = |\boldsymbol{b}^\alpha|$ and resolved shear stress τ^α is,

$$\tau^\alpha = s_i^\alpha \left(\hat{\sigma}_{ij} + \sum_{\substack{\beta=1 \\ \beta \neq \alpha}}^{n} \sigma_{ij}^\beta \right) m_j^\alpha \quad (\alpha = 1, 2, ..., N), \tag{2.14}$$

where $\boldsymbol{s}^\alpha = \boldsymbol{b}^\alpha / b^\alpha$. The velocity of dislocations v^α is assumed to be expressed by,

$$v^\alpha = \mathcal{F}(\tau^\alpha, b^\alpha), \tag{2.15}$$

where, the inertia term is neglected. As a most simple form, it is often expressed by,

$$v^\alpha = \frac{1}{B} \tau^\alpha b^\alpha, \tag{2.16}$$

where B is a resistance coefficient.

A more proper expression can be obtained, for example, by MD simulation [8]. Figure 2.2 shows the relationship between shear stress and velocity of a dislocation whose Burgers' vector is $\boldsymbol{b} = \langle 111 \rangle / 2$ on $\{112\}$ plane in bcc iron [9, 10]. It is obtained by the simple shear testing of the atomic system involving a dislocation by molecular dynamics simulation. The atomic interactions are described by the N body potential proposed by Finnis and Sinclair [11]. This result supports the linearity between v and τ at low stress levels.

2.6 Other Elementary Process of Dislocation

In addition to the relationship between shear stress and velocity of dislocations which has been mentioned in the previous subsection, various elemen-

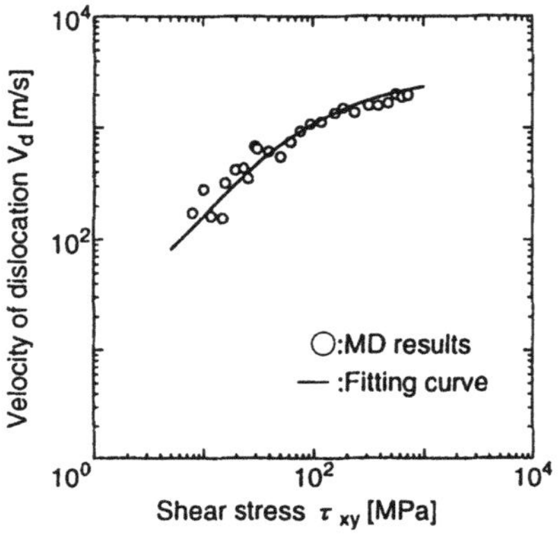

Fig. 2.2. Relationships between applied shear stress and velocity of dislocation.

tary processes related to dislocation behavior, which are listed as follows, should be taken into account;

- Peierls' force of dislocation movement;
- Mass of dislocation (consideration of inertia);
- Pinning of dislocation at grain boundary or precipitate;
- Interaction of dislocations at short range [12];
- Nucleation of dislocation from source;
- Nucleation of dislocation from crack tip and surface;
- Climb of dislocation, etc.

These processes can be taken into account in DDD simulations, if their properties are described properly. We can obtain a number of experimental data, but they contain various complex factors which cannot be separated. In this paper, these elementary processes are not taken into account for simplicity.

3. Model of Analysis

The model analyzed has a mode I crack of $a/W = 0.5$ and $L/W = 3$ with a round tip ($r_0/W = 7.5 \times 10^{-12}$) as shown in Fig. 3.1a,b.

Figure 3.1c,d indicate the mesh of FEM used for calculation of the problem B. The stress intensity factor for no dislocation included is K_{I0} / $(\sigma\sqrt{\pi a})$ = 2.824 which is evaluated by the stress extrapolating method [13] being applied to FEM results, and this value almost agrees with the analytical solution (2.826).

The material is assumed to be bcc iron and the crack front direction is set along $[1\bar{1}0]$. In this situation, two kinds of dislocations, both of edge dislocations, which have $\{112\}$ slip planes and $\langle 111\rangle$ slip directions is $(\bar{1}\bar{1}2)$, $[111]$ and (112), $[11\bar{1}]$ can be moved in the xy plane.

Young modulus, Poisson's ratio, the resistance coefficient, and the length of Burgers' vector are $E = 200$ GPa, $\nu = 0.33$, $B = 1.54 \times 10^{-5}$ Pa· s, $b/W = 6.25 \times 10^{-5}$, respectively.

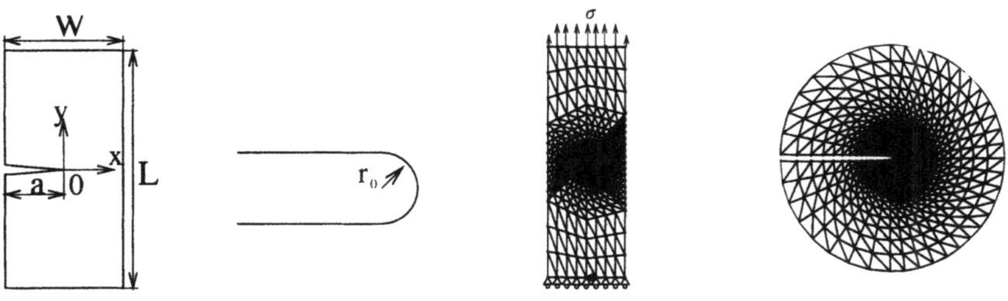

Fig. 3.1. Model of analysis. (**a**) Shape and dimensions, (**b**) detail of crack tip, (**c**) FEM mesh (all region), (**d**) FEM mesh (near crack tip).

3.1 Pile-up of Dislocations (Case 1)

The crack surface is set on (110) and pile-up of emitted edge dislocations from the crack tip which are characterized by (112) plane $[11\bar{1}]$ direction, and $\boldsymbol{b}/b = (\cos(54.7°), \sin(54.7°))$ is analyzed.

The first dislocation ($\alpha = 1$) is fixed at the point $(r^1/b, \theta^1) = (2.3 \times 10^3, 54.7°)$. $N - 1$ dislocations are set initially on the same slip plane between the first dislocation and the crack tip, and the relaxation calculations are carried out under mode I loading. N is changed from 1 to 8.

3.2 Random Distribution of Dislocation (Case 2: 1 Slip System)

The crack surface is set on (001) and 60 edge dislocations whose Burgers' vectors are parallel to $[11\bar{1}]$ and $[\bar{1}\bar{1}1]$ direction, i.e. $\boldsymbol{b}/b = (\pm\cos(35.3°), \mp\sin(35.3°))$ are initially put in the region $y > -(\tan(35.3°))x$. After relaxation without loading, the static equilibrium state is analyzed under mode I loading.

3.3 Random Distribution of Dislocation (Case 3: 2 Slip Systems)

300 edge dislocations are placed at random near the crack tip as shown in Fig. 3.2c. The crack surfaces are set on (110) and (001), which will be called by the angle θ between the slip direction and the direction of crack

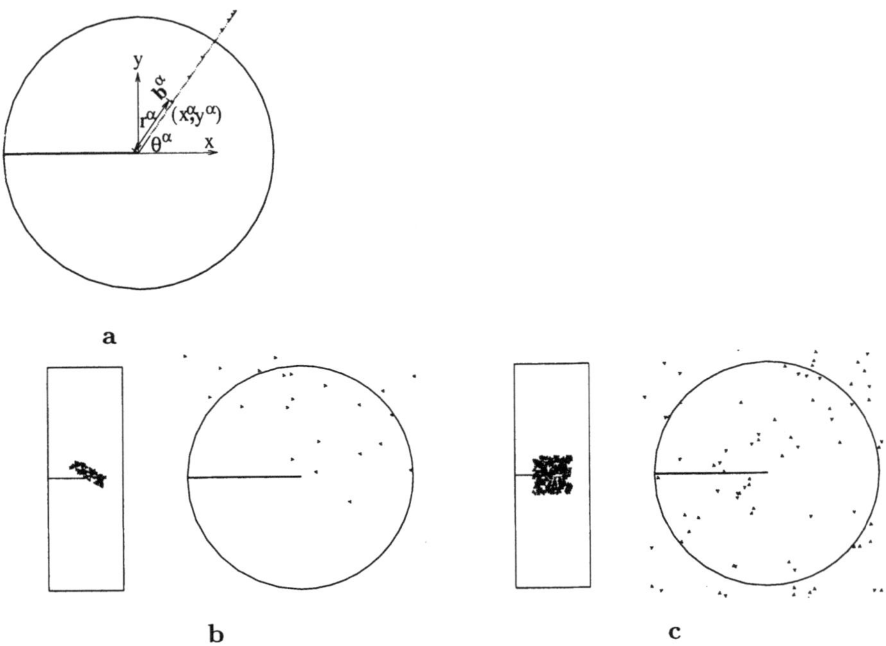

Fig. 3.2. Dislocation configuration. (**a**) Path of 8 dislocations (case 1), (**b**) initial (case 2), (**c**) initial (case 3).

propagation (x direction) as $\theta = 54.7°$, and $\theta = 35.3°$, respectively. After relaxation without loading, the succeeded relaxation is performed under mode I loading.

4. Results of Analyses

4.1 Pile-up of Dislocations (Case 1)

Figure 4.1a shows the arrangement of dislocations and equivalent stress which are obtained by the DDD simulation in case 1 ($N = 8$).

Symmetry of stress distribution with respect to x axis near the crack tip is disturbed by the effect of piled-up dislocations. Figure 4.1b shows the relationship between stress intensity factor and the number of dislocations. When the number of dislocations increases, the value of the stress intensity factor $K_{\rm I}$ decreases and a component of mode II appears. Piled-up dislocations are shown to shield the stress field in the near tip region.

4.2 Random Distribution of Dislocation (Case 2: 1 Slip System)

Figure 4.2a shows the distribution of dislocations and equivalent stress for the load free state in case 2. Pairs of dislocations which have opposite sign are arranged at 45° with respect to their Burgers' vectors relatively to each other and dislocations which have the same Burgers' vectors are arranged on a straight line. These dislocation structures build a dislocation wall in the direction of $\theta = 54.7°$ from the crack tip.

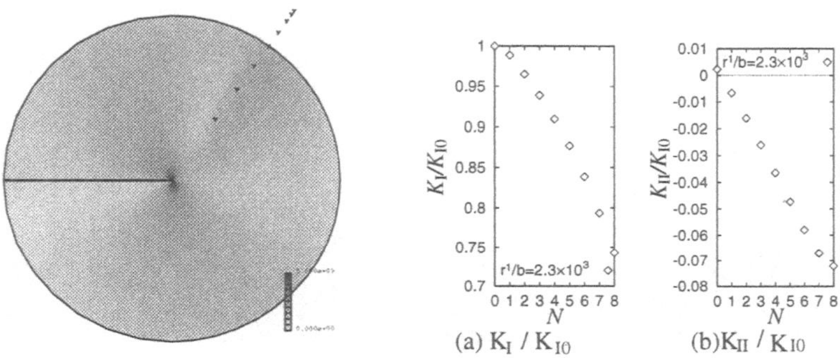

a **b**

Fig. 4.1. Result of case 1. (**a**) Equivalent stress, (**b**) stress intensity factor.

Figure 4.2b shows the path of dislocations under mode I loading. In this figure, it is observed that pairs of dislocations which have opposite signs are separated from each other, and two dislocation walls are formed by the dislocations with the same sign. Figure 4.2c,d show equivalent stress and equivalent plastic strain which are calculated by using the gradient of relative displacement evaluated by the frequency of dislocations passing through an element. The high plastic strain region expands in the direction of $\theta \approx 45°$. The stress intensity factors after loading are obtained as $K_I/K_{I0} = 1.241$ and $K_{II}/K_{I0} = 0.3292$. In this case, amplifying of the stress intensity factor is confirmed.

4.3 Random Distribution of Dislocation (Case 3: 2 Slip systems)

Figure 4.3a,b show the path of dislocations and equivalent plastic strain in the case of $\theta = 54.7°$ in case 3 under mode I loading, in which the initial arrangement is obtained by relaxation without loading. The size of the displayed region is $R/W = 0.25$. Figure 4.3c,d show the results in the case of $\theta = 35.3°$. Interactions between dislocations which belong to different slip systems form the immobile dislocations anywhere. The high plastic strain region expands in front of the crack tip upwards from the 45° to the 60°

direction. This result agrees qualitatively with the prediction obtained by crystal plasticity.

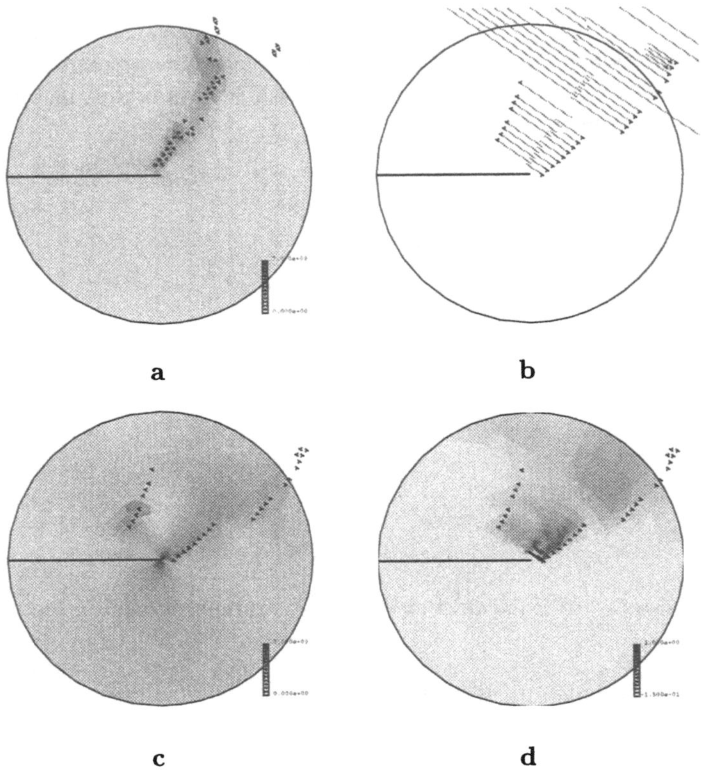

Fig. 4.2. Results of case 2. (**a**) Relaxed arrangement, (**b**) paths of dislocation, (**c**) equivalent stress, (**d**) equivalent plastic strain.

5. Conclusion

The distribution of dislocations formed under interactions between the crack tip and the dislocation field is obtained by discrete dislocation dynamics. The shielding and amplifying effects of the stress intensity factor are shown by investigation of the stress distribution.

In the discrete dislocation dynamics, the analysis is achieved by skillful separation of the problem to micro- and macroscopic analysis. In the microscopic region, which is not handled by conventional continuum mechanics, the elementary processes concerned with dislocation movement are analyzed in detail by molecular dynamics. On the other hand, in the macroscopic region, the movement of dislocations is treated dynamically in the stressed field induced by distinct dislocation and boundary conditions.

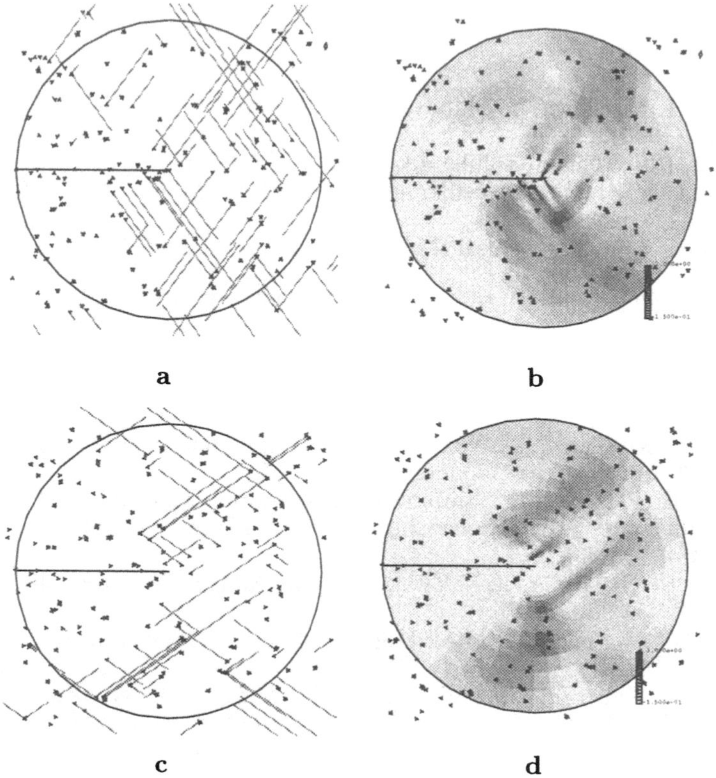

Fig. 4.3. Results of case 3. (**a**) Paths of dislocation ($\theta = 54.7°$), (**b**) equivalent plastic strain ($\theta = 54.7°$), (**c**) paths of dislocation ($\theta = 35.3°$), (**d**) equivalent plastic strain ($\theta = 35.3°$).

It is expected that this method can be applied to research of the formation process of dislocation free zones and the stress relaxation mechanism near the crack tip field. Moreover, the semi-phenomenological assumption of the hardening property of crystalline material in crystal plasticity can be revealed by these simulations.

Acknowledgement

Support for this research by a Scientific Research Grant from the Japanese Ministry of Education and Watanabe Memorial Scholarship is gratefully acknowledged.

References

1 . A. Nakatani, H. Kitagawa, Y. Shibutani, K. Nakatani: Mater. Sci. Res. Int. **1**, 11-16 (1995).
2 . S. M. Ohr: J. Phys. Chem. Solids, **48**, 1007–1014 (1987).
3 . V. Lubarda, J. A. Blume, A.Needleman: Acta Metall. Mater. **41**, 625–642 (1993).
4 . E. Vander Giessen, A. Needleman: Model. Simul. Mater. Eng. **3** 689–735 (1995).
5 . H.Kitagawa, A.Nakatani: JSME Int. J. Ser. A, **38**, 1–8 (1995).
6 . H.Kitagawa, A.Nakatani, Y.Shibutani: Mater. Sci. Eng. A **176**, 263–269 (1994).
7 . S. Suzuki: Introduction of dislocation theory, (Agne Tokyo, 1967), (in Japanese).
8 . K. Tanaka, H. Kitagawa, A. Nakatani: Prep. Jpn. Soc. Mech. Eng., No.940–39, 73–78 (1994) (in Japanese).
9 . A. Nakatani, H. Kitagawa, Y. Shibutani, K.Nakatani: Mechanical Behaviour of Materials - VII (Proceedings of the 7th International Conference), 460–461 (1995).
10. A. Nakatani, H. Kitagawa, K.Nakatani: J. Soc. Mater. Sci. Jpn., **46**, 619–624 (1997) (in Japanese).
11. M. W. Finnis, J. E. Sinclair: Philos. Mag. A **50**, 45–55 (1994) (Erratum ibid, **53**, 161 (1986)).
12. H.Kitagawa, A. Nakatani, Y. Shibutani: Proceedings of International Seminar on Microstructures and Mechanical Properties of New Engineering Materials (IMMM'93), (eds. M. Tokuda and M. Senoo), Mie University, 3–5, August, 1993, (1993) 19–26, Mie Academic Press.
13. M. Shiratori, T. Miyoshi, H. Matsushita: Numerical fracture mechanics, (Jikkyoshuppan, 1980), (In Japanese).

A Molecular Dynamics Study on the High Temperature Deformation of Zirconia Polycrystals

Hiroshi Ogawa

National Industrial Research Institute of Nagoya, AIST, 1-1, Hirate-cho, Kita-ku, Nagoya, 462 Japan.
Atom Technology Group, National Institute for Advanced Interdisciplinary Research, AIST, 1-1-4, Higashi, Tsukuba, 305 Japan.

Summary. The molecular dynamics method was applied to yttria-stabilized zirconia (YSZ) polycrystals to investigate the microscopic dynamics of grain boundaries. We analyzed the zirconia polycrystals composed of 4 grains and simulated the static and dynamic behaviors at elevated temperatures. Complex processes of the phase transition in both cooling and heating procedures were reproduced. A disordered layer at the grain boundary was observed with an estimated thickness of 1 nm. The lattice and boundary diffusion coefficients were estimated and found to have different compositional dependencies. The deformation process of polycrystals under tensile stress was also simulated.

Key Words. Polycrystal, zirconia, molecular dynamics, diffusion, deformation

1. Introduction

Mechanical properties of materials consisting of very small crystal grains are known to show some particular behaviors. One of the typical examples is the superplastic deformation of metals and ceramics [1]. Large deformation up to several hundred per cent without fracture can take place at elevated temperature. Superplastic deformation has been considered to be attained by the rearrangement of the grain configuration in the material. Ashby and Verrall [2] proposed the grain switching model where the slip of grains at the boundary is the main mechanism of the rearrangement. Although the boundary slip is a quite acceptable concept, some additional processes, such as the lattice diffusion and the boundary diffusion shown in Fig. 1, also play important roles during deformation in order to maintain the bulk continuity. The molecular dynamics (MD) method appears to be useful for investigating such microscopic transport phenomena since experimental data are quite difficult to obtain, especially at high temperatures.

The usefulness of MD simulation has been discussed for many years in the cases of molecules, single crystals, amorphous materials and liquids. Simulation of polycrystalline materials, however, frequently includes a difficulty because

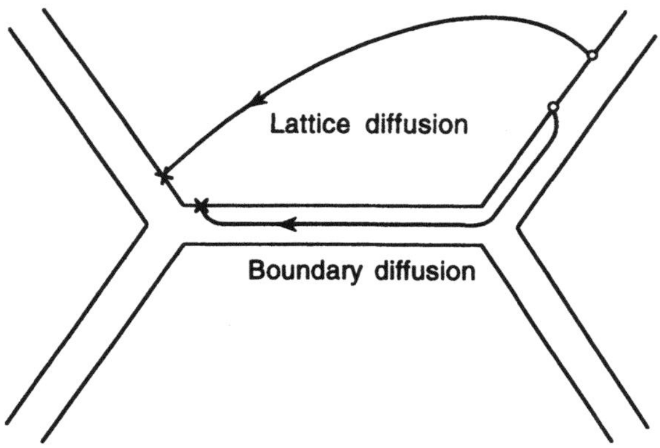

Fig. 1. Schematic diagram of the lattice diffusion and the boundary diffusion in a polycrystal.

simulation is time consuming even with a large computer system. Hence, a fully three-dimensional simulation of polycrystalline materials has been carried out only very recently [3–6]. Phillpot *et al.* [3] simulated Lennard–Jones nanocrystals composed of 8 grains using 55 296 atoms. Zhu and Averback [4] simulated the sintering process of fcc copper polycrystals. These studies dealt with a very simple grain configuration with constant grain radii. Ogawa *et al.* [5] dealt with a more complicated grain boundary structure with different grain radii. Kalia *et al.* [6] simulated a large polycrystal of silicon nitride by using parallel computers.

The purpose of the present paper is to overview our previous studies concerning the ZrO_2 polycrystal [5, 7, 8] and to show new results especially on the yttria-stabilized zirconia (YSZ) system. The author analyzed the static structure of grain boundary, lattice and boundary diffusions, and the deformation process of the model polycrystal under tensile stress.

2. MD Simulation of Polycrystals

It is not straightforward to generate a realistic, three-dimensional structure of a polycrystalline material. Unfortunately, most of the polycrystalline structures previously used in MD simulation are quite simple, such as the stacking of cubic grains. Recently the present author proposed a convenient way for expressing more complex polycrystalline structures [5]. In the model, the MD basic cell is filled by N grains of shapes which are expressed by the weighted Voronoi polyhedra

V given by,

$$V_i = \{ \boldsymbol{r} \mid d(\boldsymbol{r}, \boldsymbol{P}_i) < d(\boldsymbol{r}, \boldsymbol{P}_j),\ j \neq i \}, \quad 1 < i < N, \tag{1}$$

$$d(\boldsymbol{r}, \boldsymbol{P}_i) = \mid \boldsymbol{r} - \boldsymbol{P}_i \mid^2 - w_i^2, \tag{2}$$

where $\boldsymbol{P}_i$ and w_i are the center position and radius of i-th grain. The geometry of the weighted Voronoi polyhedron appears to describe the configuration of grains in polycrystals well [5]. Beside the $4N$ geometrical parameters, $\boldsymbol{P}_i$ and w_i, further $6N$ parameters are required to describe the orientation of crystal axes and the translational shift in each grain region. Hence, the model structure is uniquely defined by using $10N$ geometrical parameters.

The polycrystalline structure used in this work was composed of four grains as shown in Fig. 2. The size of the MD basic cell is approximately 8 nm × 8 nm × 3 nm, and the periodic boundary conditions were applied to the system in the x, y and z directions. The shapes of the grains are infinitely long toward the z axis, and the approximate radii of larger and smaller grains projected onto the xy plane are 5 nm and 3 nm, respectively. The grain regions are filled by atoms by changing the axis orientation which satisfy the continuity toward the z direction. The values of the geometrical parameters are the same as those in the previous study [5]. The target compositions are pure ZrO_2 and $0.25\ YO_{3/2} - 0.75\ ZrO_2$. In the latter case, substitutions of Y and oxygen vacancy for Zr and O, respectively, are carried out randomly. The number of atoms is about 12000 in each composition.

In the MD simulation, the equation of motion was integrated by using the Verlet algorithm [9] with a time step of 2 fs. Coulombic potentials of the Born-Mayer

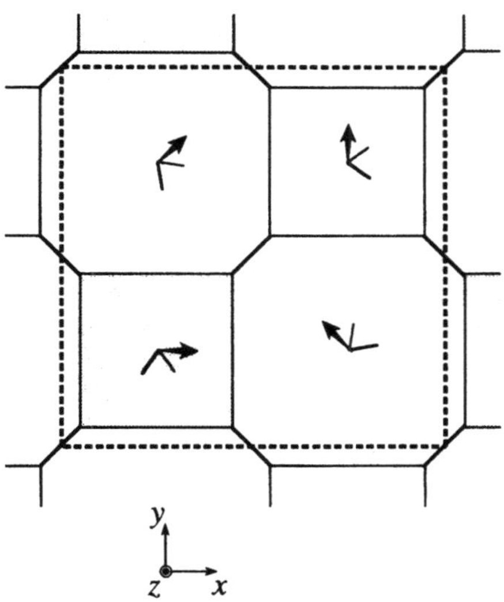

Fig. 2. The arrangement of grains and their boundaries in the present model projected onto the xy plane. The arrow and the accompanied lines in each grain denote the orientation of the crystal axes. The MD basic cell is shown by the dotted rectangle.

type proposed by Dwivedi and Cormack [10, 11] were used. This potential model was constructed based on the energy surfaces of cubic, tetragonal, and monoclinic structures of zirconia. The original form of the potential includes the rigid ion part and the shell term. We ignored the shell term in the present simulation due to the limited computer resources. The force calculation was optimized by the neighbors-list and the Ewald sum methods [9]. The temperature of the system was controlled by a Nosé thermostat [12], and the heating and cooling rates of $\pm 25\,K\,ps^{-1}$ were used to evaluate the temperature dependencies. The diagonal parts of the pressure tensor were adjusted to 0.1 MPa using Andersen's method [13] for each direction.

3. Results

3.1 Static Structures

The stability of the model polycrystal at elevated temperatures was examined for pure ZrO_2. In Fig. 3 is shown a snapshot of the simulated sample at 2500 K after holding the system at atmospheric pressure for 80 ps. The simulated density of the polycrystal is $5.48\,Mg\,m^{-3}$ at 2500 K which is fairly consistent with experimental values for cubic zirconia [14]. At this temperature, the simulated sample keeps the original configuration of the grain boundary. At 3000 K which is just above the melting point of ZrO_2, ~ 2960 K [15], on the other hand, the thickness of the boundary layer increases and the two smaller grains were melted. Although the precise value of the melting point cannot be estimated by the ordinary MD method, it is apparent that the polycrystalline structure is replicated by the simulation even at temperatures near the melting point.

It is well known that zirconia shows complex phase transitions between cubic, tetragonal, and monoclinic phases. In the present simulation, similar phase transition phenomena were partly reproduced. In Fig. 4 the density variations of the simulated sample on cooling and heating are shown. Points where the phase transitions occurred are denoted by arrows in the figure. It is interesting that the transition does not always occur for all grains simultaneously. After the transition, crystals were elongated toward the *c* axis and the oxygen atoms were distributed on two sites. By averaging the atomic coordination inside the grains, the positional parameter of oxygen, z_O, was evaluated to be about 0.17, which is fairly close to the experimental value of the tetragonal phase, 0.185 [16]. The monoclinic phase was not reproduced in the present simulation. Hysteresis of the density variation on cooling and heating was also observed, as shown in Fig. 4. Such hysteresis is considered to be caused by the large gradient of temperature control which is inevitable in the MD simulation.

The disordered layers at the grain boundary were recognized in all simulated samples below the melting point. In order to evaluate the thickness of the boundary

Fig. 3. Snapshot of the simulated structures as projected onto the *xy* plane at 2500 K. Oxygen and zirconium atoms are denoted by gray and black spheres, respectively. Size of the displayed region is about 12 nm × 12 nm.

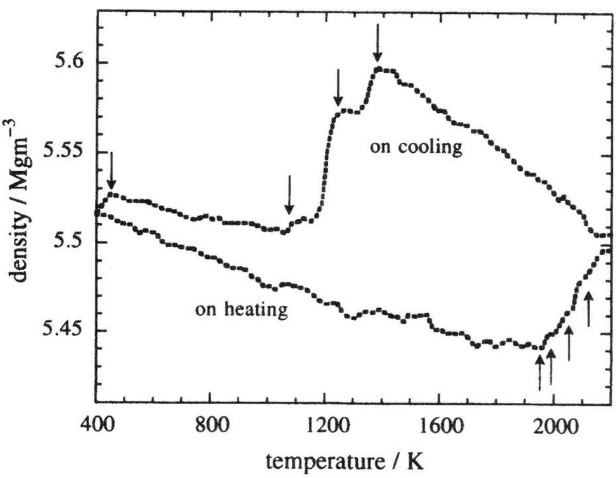

Fig. 4. Density variation during the cooling and heating processes. The arrows denote the points where the phase transitions take place.

layer, potential energy values of all atoms were calculated and mapped out on the *xy* plane as shown in Fig. 5. In the figure, bright regions are energetically unstable compared with the dark regions inside the grains. The thickness of the boundary region was estimated to be roughly 1 nm. This result is comparable to the few nm boundary layer thickness estimated from the TEM observation of yttria-stabilized zirconia [17]. Although we should note that the boundary layers in the real material

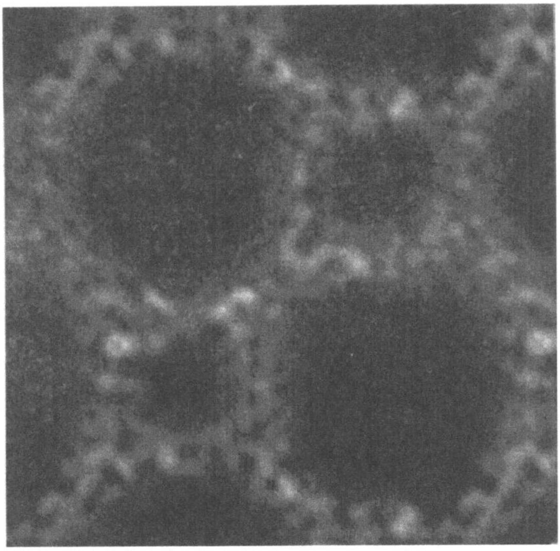

Fig. 5. Potential energy distribution in the simulated sample as the projection onto the *xy* plane. The displayed region is about 10 nm × 10 nm.

are formed by impurities such as silicates, the similarity between the results of simulation and experiments is remarkable.

3.2 Lattice Diffusion and Boundary Diffusion

Zirconia ceramics are known as super-ionic conductors and are widely used as oxygen sensors. Many MD studies have been carried out to investigate the self-diffusion of oxide ions for single crystal cases [10, 11, 18]. Diffusion processes of oxide ions in polycrystals are important not only for sensors but also in relation to the superplastic deformation as mentioned above. In Fig. 6 the trajectories of oxide ions in the simulated sample during 10 ps are shown. It is clear that the diffusion processes inside the grain and at the boundary layer are very different: oxide ions move toward <100> directions inside the grain, but travel irregularly at the grain boundary region.

In order to evaluate the diffusion coefficients concerning the lattice and boundary diffusions, all atoms in the MD basic cell are divided initially into two groups: atoms inside the grain and those at the boundary layer. The thickness of the boundary layer was assumed to be 0.5 nm. The atoms in the transient region between 0.25 to 1 nm from the boundary were excluded from both groups so as to avoid unnecessary confusion. The self-diffusion coefficients in each region were calculated by the linear regressions of mean square displacements based on the trajectory data during 60 ps. The resulting values for pure ZrO_2 are 1×10^{-11} and

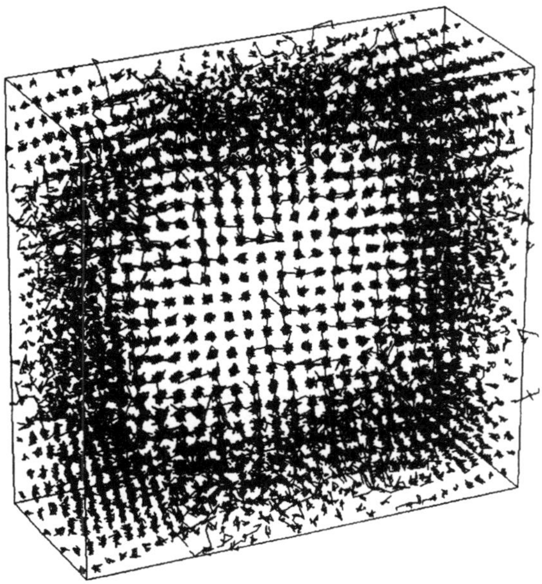

Fig. 6. Trajectories of the oxide ions during 10 ps at 2000K. The rectangular region of about 6 nm × 6 nm × 1 nm is projected onto the <001> direction of the central crystal region.

5×10^{-11} $m^2 s^{-1}$ for lattice and boundary diffusions, respectively. The diffusivity inside the grain is relatively low because few oxygen vacancies exists in pure ZrO_2 crystals.

Diffusivities of 0.25 $YO_{3/2}$ – 0.75 ZrO_2, on the other hand, are relatively large due to the oxygen vacancies. The simulated values are 7×10^{-10} and 3×10^{-10} $m^2 s^{-1}$ for lattice and boundary diffusions respectively, and the former value is fairly comparable to the experimental value [19]. It is interesting that the diffusivity at the boundary is lower than inside grains at this composition which is the reverse of that in the pure ZrO_2 case. This information is expected to be useful for understanding the superplastic properties of actual samples more precisely.

3.3 Deformation under Tensile Stress

Simulation of the deformation process under tensile stress was carried out for the pure ZrO_2 polycrystal. We assumed the tensile stresses of 4, 6 and 8 GPa while holding the temperature at 1500 K. The simulated stress–strain curves are plotted in Fig. 7. At 4 GPa tensile stress, the sample elongates about 3% but then remains stationary for more than 10 000 MD steps. Plastic deformations after reaching the elastic limit were observed at tensile stresses greater than 6 GPa, and a large plastic deformation of more than 20% was observed at 8 GPa.

In Fig. 8 a snapshot of such an elongated sample at 8 GPa is shown. Small pores at the grain boundary are recognized, and the thickness of the boundary layer

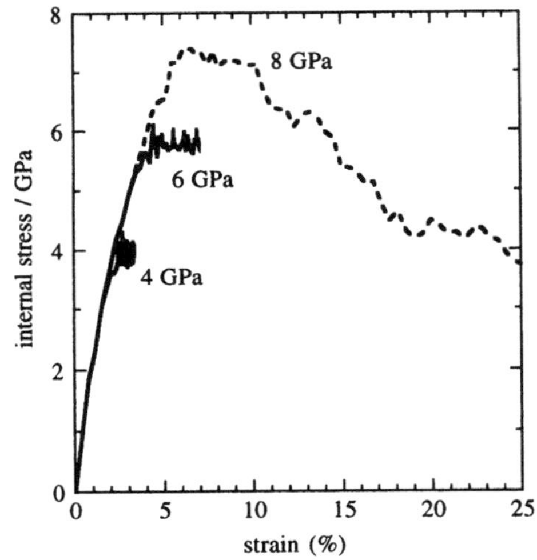

Fig. 7. Simulated stress-strain curves at 1500 K of the model ZrO_2 polycrystal. Applied tensile stresses are 4, 6 and 8 GPa.

seems to increase compared with the initial structure. Variations of the crystal structure were found in some grains, and the original image is almost destroyed in the case of the lower left grain. As far as we observe in the present simulation, deformation of the sample is attained mainly by the variation of grain shapes, and the rearrangement of the grain configuration, such as the grain switching, was not clearly observed.

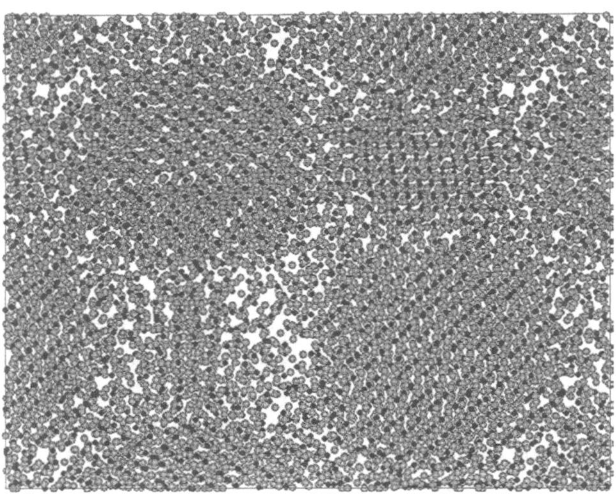

Fig. 8. Deformed structure of the present model by 8 GPa tensile stress at 1500 K. The strain is about 20%.

4. Conclusion

The present MD simulation extracted several interesting features of YSZ polycrystals on both static and dynamical aspects as shown above. Although the simulated grain boundary configuration is simple compared with actual materials, simulated properties concerning the phase transitions and diffusion are informative. It would be preferable, however, to take impurity elements into account in order to reproduce more precisely the characteristics of grain boundaries.

In conclusion, the author stresses that MD simulation appears to be useful for investigating the deformation phenomena of polycrystalline materials at high temperatures. It would be interesting to extend the present method to various grain configurations to test its validity in a wider base.

Acknowledgement

The MD simulation was carried out on a CRAY-C90 at the RIPS center, AIST, Tsukuba, and at the JRCAT Supercomputer System, Tsukuba.

References

1. T.G. Nieh, J. Wadsworth, O.D. Sherby: *Superplasticity in metals and ceramics,* (Cambridge University Press, 1997).
2. M.F. Ashby, R.A. Verrall: Acta Metal. **21**, 149 (1973).
3. S.R. Phillpot, D. Wolf, H. Gleiter: J. Appl. Phys. **78**, 847 (1995).
4. H. Zhu, R.S. Averback: Mater. Sci. Eng. A **204**, 96 (1995).
5. H. Ogawa, F. Wakai, Y. Waseda: Mol. Simulat. **17**, 179 (1996).
6. R.K. Kalia, A. Nakano, K. Tsuruta, P. Vashishta: Phys. Rev. Lett. **78**, 689 (1997).
7. H. Ogawa, N. Sawaguchi, F. Wakai: Trans. MRSJ **20**, 807 (1996).
8. H. Ogawa, N. Sawaguchi, F. Wakai: Mater. Sci. Forum **243–245**, 351 (1997).
9. M.P. Allen, D.J. Tildesley: *Computer Simulation of Liquids*, (Oxford, 1987).
10. A. Dwivedi, A.N. Cormack: J. Solid State Chem. **79**, 218 (1989).
11. A. Dwivedi, A.N. Cormack: Philos. Mag. A **61**, 1 (1990).
12. S. Nosé: J. Chem. Phys. **81**, 511 (1984).
13. H.C. Andersen: J. Chem. Phys. **71**, 2384 (1980).
14. R.P. Ingel, D. Lewis III: J. Am. Ceram. Soc. **69**, 325 (1986).
15. V.S. Stubican: in "Advances in Ceramics Vol. 24: Science and Technology of Zirconia III", (The American Ceramic Society, 1988) p.71.

16. G. Teufer: Acta Crystallogr. **15**, 1187 (1962).
17. M. Rühle, N. Claussen, A.H. Heuer: in "Advances in Ceramics Vol. 24: Science and Technology of Zirconia III", (The American Ceramic Society, 1988) p.352.
18. X. Li, B. Hafskjold: J. Phys. Condens. Matter **7**, 1255 (1995).
19. Y. Oishi, K. Ando: in "Transport in Nonstoichiometric compounds", eds. G. Simkovich and V.S. Stubican, (Plenum Press, New York, 1985) p.189.

Part 5

Electronic Structure Calculations

Ab Initio Calculations of SiC Grain Boundaries and SiC/Al Interfaces

Masanori Kohyama and John Hoekstra

Department of Material Physics, Osaka National Research Institute, AIST, 1-8-31, Midorigaoka, Ikeda, Osaka 563, Japan

Summary. *Ab initio* calculations of SiC grain boundaries and SiC/Al interfaces have been performed using the techniques of the first-principles molecular dynamics method. For the Σ=9 boundary in SiC, the interfacial C–C and Si–Si bonds have siginificant effects by generating peculiar bond lengths, bond charges and localized states. For the SiC(001)/Al interface, the C-terminated and Si-terminated interfaces have quite different features from each other, such as atomic configurations, bonding nature, bond adhesion and electronic properties.

Key Words. *Ab initio* pseudopotential method, density-functional theory, grain boundary, SiC, metal/ceramic interface

1. Introduction

Grain boundaries dominate various properties of ceramics, especially mechanical ones. Moreover, it is necessary to form metal/ceramic interfaces in their practical applications. Thus, it is of great importance to investigate grain boundaries in ceramics and metal/ceramic interfaces, where an understanding of atomic bonding is essential. Currently, by virtue of the development of the first-principles molecular dynamics (FPMD) method [1–4], it is possible to apply *ab initio* calculations based on the density-functional theory [5, 6] to such complex systems. In this paper, we present our recent *ab initio* calculations of grain boundaries in SiC and SiC/Al interfaces by using the techniques of the FPMD method applied to interfaces [7].

As for grain boundaries in covalent ceramics, there exist only a few atomistic calculations, in contrast to various calculations of grain boundaries in elemental semiconductors [8]. Recent studies of coincidence-site-lattice (CSL) tilt boundaries in Si have shown that such interfaces are constructed by structural units consisting of atomic rings without any dangling bonds [8]. In the case of SiC, there should exist further complexities such as polar and non-polar interfaces and interfacial wrong bonds between like atoms. Our previous tight-binding (TB) calculations of the $\{122\}\Sigma$=9 tilt boundary in SiC [9, 10] indicated that such wrong bonds have significant effects. In this paper, we deal with the same boundary in SiC by using the *ab initio* method.

For metal/ceramic interfaces, it is of great importance to understand the bonding nature and the origins of adhesion [11, 12]. For non-reactive interfaces formed by ionic ceramics and metals, recent *ab initio* calculations [12, 13] have shown that the origin of adhesion is mainly explained by electrostatic effects such as the image interaction and the van der Waals interaction. For reactive interfaces formed by covalent ceramics and metals, there should exist rather strong interactions such as charge transfer and orbital hybridization. In our previous *ab initio* calculations of the SiC(110)/Al interface [7], strong attractive C–Al interactions with both covalent and ionic characters have been found. In this paper, we deal with the SiC(001)/Al interface. Here, the differences between the C–Al and Si–Al interactions can be investigated clearly by examining the C-terminated and Si-terminated interfaces.

2. Theoretical Method

The electronic ground state of a large system can be efficiently obtained by the FPMD method in the framework of the plane-wave pseudopotential method. Thus, relaxations or molecular dynamics, according to the exact Hellmann–Feynman forces, can be easily performed. Various kinds of schemes for obtaining the electronic ground state have been developed since the original Car–Parrinello algorithm [1–4]. The principle of such schemes can be explained as follows. The purpose is to obtain a set of occupied wave functions $\{\psi_i\}$ which minimize the Kohn–Sham total-energy functional $E_{tot}[\{\psi_i\}, \{\mathbf{R}_n\}]$ for given atomic positions $\{\mathbf{R}_n\}$. For the minimization of E_{tot} with the orthonormalization constraint of $\{\psi_i\}$, we consider the functional $E = E_{tot} - \sum \lambda_{ij}(\langle\psi_i|\psi_j\rangle - \delta_{ij})$ where λ_{ij} are the Lagrange multipliers. Its gradient with respect to ψ_i can be expressed as $\delta E/\delta\psi_i^* = (H - \lambda_i)\psi_i$ where $\lambda_i = \langle\psi_i\,|H|\,\psi_i\rangle$, if the orthonormalization process is performed separately. The negative of this gradient can be regarded as a force or a steepest-descent vector acting on ψ_i. Thus, the wave functions can be iteratively updated according to this force and can reach the true eigenfunctions minimizing E_{tot}. This iterative update is performed by using 2nd or 1st-order equations of motion [1] or by using conjugate-gradient (CG) techniques [2–4].

It is of great importance to determine the most efficient schemes for the relevant interface systems. We have performed convergence tests for long supercells of interfaces [7]. It has been shown that the band-by-band CG method by Teter, Payne and Allan (TPA) [2] or the all-band CG method [3] are suitable for non-metallic large systems such as grain boundaries in SiC. For the SiC/Al interface, the CG method by Bylander, Kleinman and Lee (BKL) [4] is the most efficient, although the other methods cannot obtain the electronic ground state because of strong charge sloshing.

Another important technique in the FPMD method is the use of soft or optimized pseudopotentials, so as to reduce the number of plane-wave basis functions. We have constructed the Troullier–Martins optimized pseudopotentials [14]. The properties of bulk Si, SiC and diamond are reproduced accurately within the local density approximation [15] by using a plane-wave cutoff energy (E_{cut}) of 35 Ry for Si and 60 Ry for SiC and diamond. The equilibrium lattice constants of Si, S iC and diamond are 98.9%, 98.9% and 99.3% of the experimental values. For grain boundaries in SiC, we use an E_{cut} of 60 Ry, because boundaries contain C–C bonds. For the SiC/Al interface, we use an E_{cut} of 45 Ry, enough to reproduce the properties of bulk SiC, where the equilibrium lattice constant is 99.3% of the experimental value.

3. Grain Boundaries in SiC

3.1 Atomic Models

We deal with the {122}Σ=9 boundary as a typical CSL boundary. Figure 3.1 shows the four-fold coordinated models of the polar and non-polar interfaces of this boundary in the zinc-blende structure [10]. The bonding network is common to the same boundary in Si [8]. These three models are constructed by inverting the polarity of grains. These are consistent with the electron microscopy observation in chemical vapour-deposited SiC [16], although the polarity was not determined. Polar and non-polar interfaces can be defined by the interface stoichiometry, and polar interfaces are non-stoichiometric. The polar interfaces contain one type of wrong bond, and the non-polar interface contains the same numbers of the two types of wrong bonds.

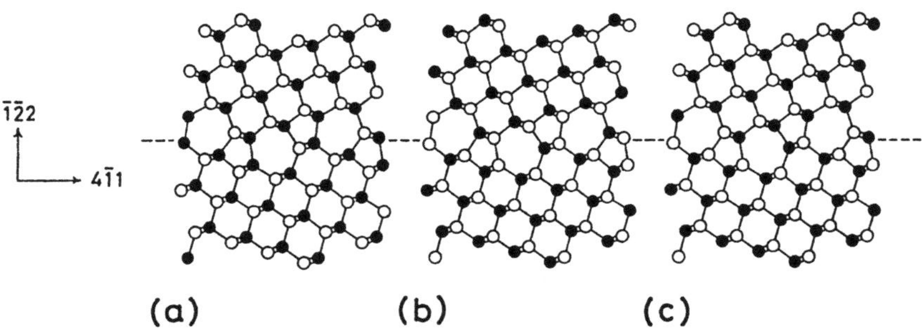

Fig. 3.1. Atomic models of the {122}Σ=9 boundary in SiC. (**a**) N-type polar, (**b**) P-type polar and (**c**) non-polar interfaces. Atomic positions are projected along the ⟨011⟩ axis. Empty and filled circles represent Si and C atoms, respectively.

We use 64-atom supercells. The supercell for the polar interfaces must contain two inequivalent interfaces alternately by the crystallographic constraint. This may cause serious problems if charge transfer occurs between the interfaces, although there exists negligible transfer in the present case. Initial configurations are constructed by introducing the rigid-body translations between bulk regions optimized by TB calculations [9, 10]. All the atoms are relaxed according to the atomic forces.

3.2 Stable Configurations

Figure 3.2 shows the final configuration of the non-polar interface. The bond lengths and bond charges of the C–C and Si–Si bonds have features similar to those in bulk diamond and Si. The C–C bond is 3.6% longer than that in bulk diamond. The Si–Si bond is 4.4% shorter than that in bulk Si. Here, the bond lengths in bulk diamond and Si mean the calculated equilibrium values. The C–C bond has a two-peak structure in charge density, typical of bulk diamond, although the Si–Si bond has a one-peak structure. The peak value of the C–C bond charge is smaller than that of bulk diamond because of stretching, although that of the Si–Si bond is larger than that of bulk Si because of compression. The bond-length and bond-angle distortions of the other bonds range from –2.9% to +2.9% and from –22.4° to +27.9°.

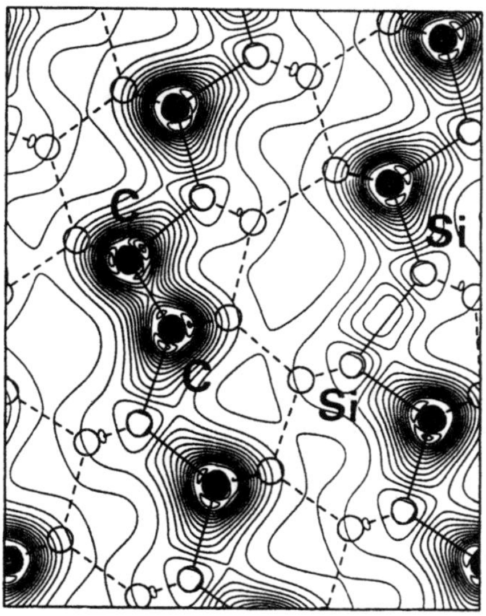

Fig. 3.2. Relaxed configuration of the non-polar interface of the Σ=9 boundary in SiC. Contours of the valence electron density are plotted from 0.015 a.u.$^{-3}$ to 0.295 a.u.$^{-3}$ with a spacing of 0.020 a.u.$^{-3}$. Circles indicate the atomic positions. Solid lines indicate bonds on the (011) plane on which the charge density is plotted. (a.u. are atomic units.)

Figure 3.3 shows the final configurations of the two types of polar interfaces. In the N-type interface, the C–C bond is 4.5% longer than that in bulk diamond, and the distortions of the other bonds range from –2.7% to +2.0% and from –20.1° to +22.5°. In the P-type interface, the Si–Si bond is 1.1% shorter than that in bulk Si, and the distortions of the other bonds range from –2.7% to +2.5% and from –13.0° to +24.0°. The bond charges of the wrong bonds have features similar to those in bulk diamond and Si. The bond distortions in the polar and non-polar interfaces are comparable to the TB results [9, 10] except for the C–C bond at the N-type interface.

The interfacial energy of the non-polar interface is 1.3 J m^{-2} from the energy increase against the value of bulk SiC. This is comparable to the TB results, 1.4 J m^{-2}. The total energy of a pair of polar interfaces is slightly lower than that of the non-polar interfaces, by about 0.1 eV per supercell.

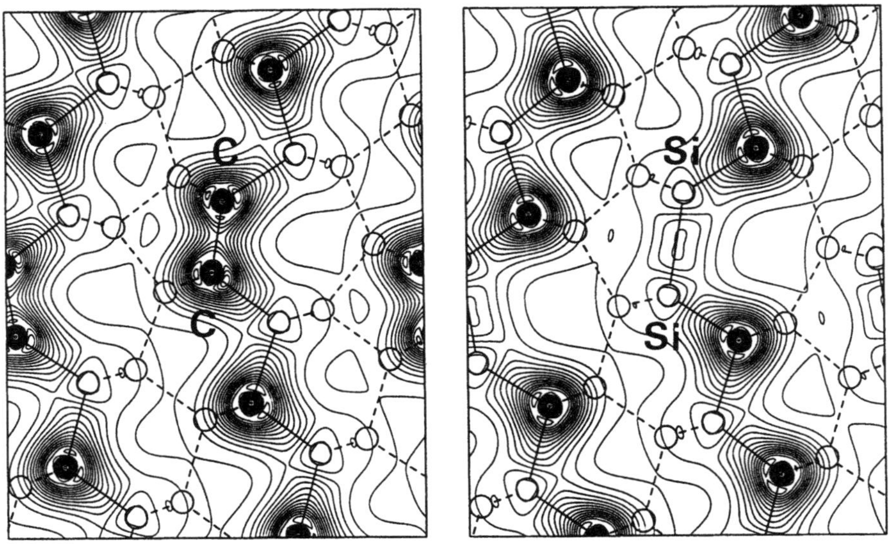

Fig. 3.3. Relaxed configurations of the N-type polar interface (*left*) and the P-type polar interface (*right*) of the Σ=9 boundary in SiC.

3.3 Electronic Structure

The electronic structures of the present interfaces have clear band gaps. This feature, as well as the bond distortions and bond charges, indicates that all the interfacial bonds are well reconstructed. We have analyzed the wave functions at Γ point. For the non-polar interface of SiC, states strongly localized at the C–C bonds are observed at the top and bottom of the valence band. The highest occupied states exist at about 0.3 eV above the upper bulk valence-band edge of SiC, although the localized states at the lower valence-band edge are close to the lower bulk band edge. Similar C–C bond localized states are also found for the N-type polar interface. States weakly localized at the Si–Si bonds are also observed near the upper valence-band edge for the non-polar and P-type polar interfaces. The C–C bond states above the upper bulk band edge should greatly affect the electronic properties of p-type samples. Such states were not found in the TB calculations [9, 10].

The origin of the C–C bond states can be explained by the analogy to a diamond cluster embedded in SiC. The width of the valence band of diamond is very much larger than that of SiC. Thus, a diamond cluster probably generates electronic states above and/or below the bulk valence-band edges of SiC, which should become localized states. The C–C bond states at the lower valence-band edge consist of s-like orbitals of C atoms, and the C–C bond states above the upper bulk valence-band edge consist of p-like orbitals. These features are common to the lower and upper valence-band edge states of bulk diamond at the Γ point. The reason why the C–C bond states severely stick out above the upper bulk band edge can be explained by the polarization of the back Si–C bonds, which should cause upward shifts of the electrostatic potential at the C–C bond.

4. SiC/Al Interfaces

4.1 Atomic Models

The C-terminated and Si-terminated interfaces of the SiC(001)/Al interface have been examined. Each supercell contains a slab of 9 SiC(001) atomic layers with 5 Al(001) layers on both surfaces. Two free Al surfaces are separated by a vacuum region of about 15 a.u., which ensures stable interlayer distances without any constraint. Coherent interfaces with ideal (1×1) periodicity are dealt with, where Al layers are expanded along the interface by about 8%, as the present purpose is to clarify t he essential features of interfacial bonds. For the Al layers on the SiC surface, there exist four special translations, corresponding to the energy extremes. These are the bridge site, the four-fold site, the on-top site over the surface atom, and the on-top site over the first atom below the surface. We performed relaxations from these initial translations, and determined the most stable configurations.

4.2 C-Terminated Interface

In the most stable configuration of the C-terminated interface shown in Fig. 4.1(a), the interfacial C atoms are three-fold coordinated. The C–Al bond length is almost the same as the back C–Si bond length, and the charge distributions around these three bonds have similar features. At the C–Al bond, the charge density is pulled from the Al atom toward the C atom, and a large charge density is concentrated near the C atom, while the charge density at the back C–Si bonds is slightly reduced from bulk levels. The charge transfer from the Al atom toward the C atom is also clear in Fig. 4.2(a). There exists a charge depletion region near the interfacial Al atom. The depletion at the interfacial C atom against the bulk C atoms is due to the reduced coordination.

Figure 4.3(a) shows the local densities of states (LDOS) of this interface. The LDOS curves at the interface and at the back Si–C bonds are substantially different from the bulk regions. This indicates that the back Si–C bonds and the interface have a peculiar electronic structure, which seems to be related to the three-fold coordination. We have analyzed the wave functions of several gap states localized at the interface region. We have found occupied states consisting of bonding p orbitals between the interfacial C and Al atoms, and unoccupied ones consisting of non-bonding p orbitals of the interfacial C atoms. These features indicate that the interactions between the sp^2 hybridized orbitals and non-bonding p orbital of the interfacial C atom and the orbitals of neighboring Al and Si atoms may stabilize the three-fold coordination, as well as the effects of charge transfer.

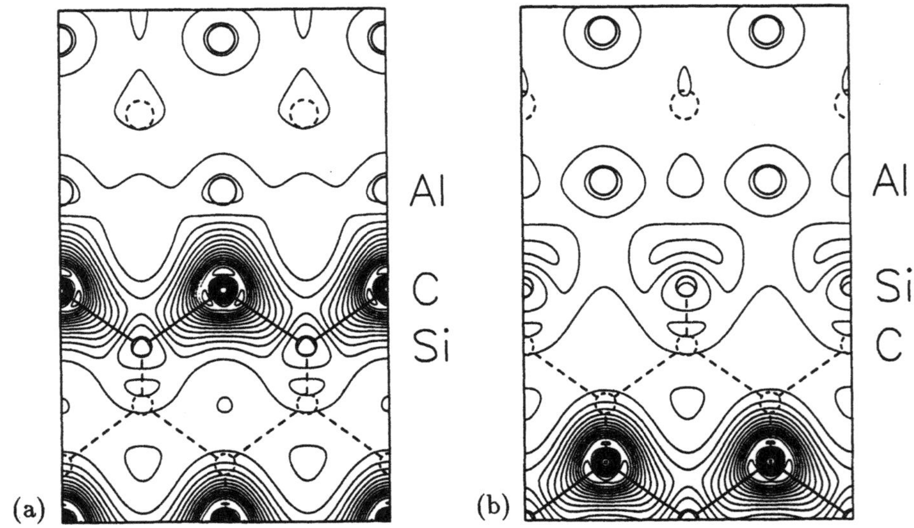

Fig. 4.1. Stable configurations of the SiC(001)/Al interface. (**a**) The C-terminated interface and (**b**) the Si-terminated interface. Contours of the valence electron density are plotted from 0.01 a.u.$^{-3}$ to 0.29 a.u.$^{-3}$ with a spacing of 0.02 a.u.$^{-3}$ on the $(1\bar{1}0)$ plane for the C-terminated interface, and from 0.01 a.u.$^{-3}$ to 0.27 a.u.$^{-3}$ in the same spacing on the (110) plane for the Si-terminated interface. Solid lines indicate atoms and Si–C bonds on the plane on which the charge density is plotted.

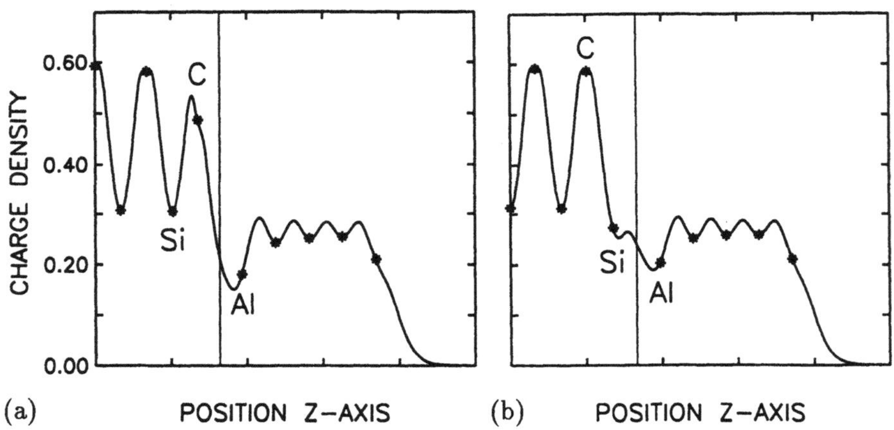

Fig. 4.2. Valence charge density distribution along the ⟨001⟩ axis in each supercell of the SiC(001)/Al interface. (**a**) The C-terminated interface and (**b**) the Si-terminated interface. Charge densities are averaged on each (001) plane.

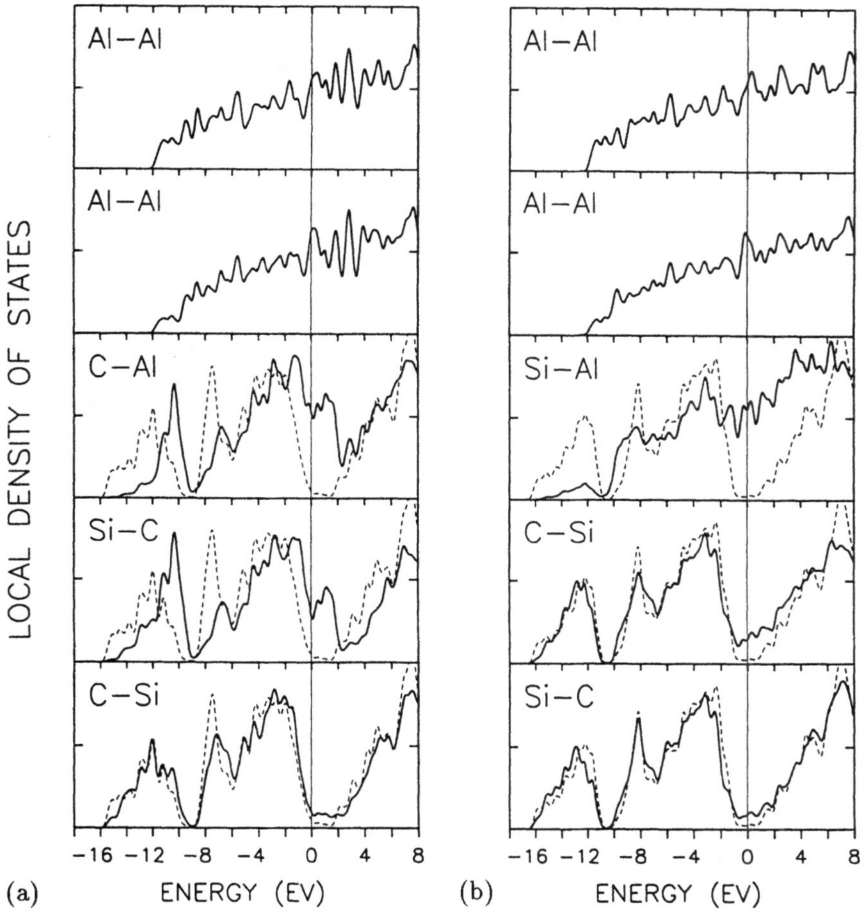

Fig. 4.3. Local densities of states (LDOS) of the SiC(001)/Al interface. (**a**) The C-terminated interface and (**b**) the Si-terminated interface. LDOS curves are calculated for each region between successive (001) atomic layers in each supercell. Broken lines indicate the LDOS of the bulk SiC region in each supercell. Vertical l ines indicate the Fermi level of each supercell.

4.3 Si-Terminated Interface

As shown in Fig. 4.1(b), the interfacial Si atoms are four-fold coordinated. The distance between the interfacial Si and Al atoms is much larger than the C–Al bond in the C-terminated interface. The charge density between the interfacial Si and Al atoms is rather broadly distributed. In Fig. 4.2(b), only a shallow depletion region is present near the interfacial Al atom, although a slight increase of charge density near the Si atom can be seen. This is in contrast to the large depletion in the C-terminated interface. In Fig. 4.3(b), the LDOS at the interface region has an appearance similar to that of the metallic Al region instead of the bulk SiC region. This also suggests a more metallic nature of the Si–Al bond, as well as the broad distribution of the charge density. The LDOS at the back C–Si bond recovers very quickly to features similar to the bulk SiC region, which is consistent with the four-fold coordination of the interfacial Si atom.

4.4 Bond Adhesion and Schottky-Barrier Height

The adhesive energy is $6.42\,\mathrm{J\,m^{-2}}$ for the C-terminated interface, and $3.74\,\mathrm{J\,m^{-2}}$ for the Si-terminated interface. The C–Al bond with covalent and ionic characters is twice as strong as the more metallic Si–Al bond. However, both values are much larger than the calculated values for non-reactive metal/ceramic interfaces, for example, $1.10\,\mathrm{J\,m^{-2}}$ for the MgO(001)/Al interface [13]. These values are also larger than the adhesive energy of two Al(001) surfaces, $2.04\,\mathrm{J\,m^{-2}}$ [13], which means that the SiC/Al interface is stronger than bulk Al.

One very important property of metal/ceramic interfaces as electronic devices is the Schottky-barrier height (SBH). The SBH can be calculated as the difference between the Fermi level of the supercell and the upper valence-band edge in the bulk SiC region, which corresponds to the SBH for p-type samples. For the C-terminated interface, the SBH is 0.08 eV, and for the Si-terminated interface, 0.85 eV. The latter is in good agreement with the experimental value, 0.9 eV, for Si-terminated interfaces [17]. The dramatically smaller SBH of the C-terminated interface can be explained by the interface dipole caused by the charge transfer from the Al layer to the C layer, which shifts the electrostatic potential at the Al layers relatively downward. The interface dipole substantially affects the band discontinuity of two materials [18]. In previous models for the SBH [19], it was believed that the SBH was intrinsic to particular semiconductors or a particular combination between metal and semiconductor. However, recent experiments and *ab initio* calculations have indicated the dependence on the interfacial structure [19]. The present results support such findings.

5. Conclusion

The present calculations have shown that the interfacial C–C and Si–Si bonds have significant effects on grain boundaries in SiC, and that the interfacial C–Al and Si–Al bonds dominate the properties of SiC/Al interfaces. These results are quite valuable for the design of SiC ceramics and SiC/Al interfaces. In addition to the calculations of stable configurations, it is possible to perform tensile or shear tests of each interface of the present systems by using the FPMD method, which will be achieved in the near future. Such calculations should clarify more realistically the effects of interfacial bonds on the mechanical properties of interfaces.

Acknowledgments

The authors are grateful to Dr. A. Sawamura and Dr. S. Ogata for fruitful discussions. M. K. than ks Dr. D.R. Hamann for sending us his computer code for constructing generalized pseudopotentials, which was modified for the construction of the TM pseudopotentials. The present study was supported by the Science and Technology Agency of Japan as the project *"Frontier Ceramics"*. All the calculatons were performed on the CRAY C90 at RIPS center of the Agency of Industrial Science and Technology of Japan.

References

1. R. Car, M. Parrinello: Phys. Rev. Lett. **55**, 2471 (1985).
2. M.C. Payne, M.P. Teter, D.C. Allan, T.A. Arias, J.D. Joannopoulos: Rev. Mod. Phys. **64**, 1045 (1992).
3. R.D. King-Smith, D. Vanderbilt: Phys. Rev. B **49**, 5828 (1994).
4. D.M. Bylander, L. Kleinman, S. Lee: Phys. Rev. B **42**, 1394 (1990).
5. P. Hohenberg, W. Kohn: Phys. Rev. **136**, B864 (1964).
6. W. Kohn, J.L. Sham: Phys. Rev. **140**, A1133 (1965).
7. M. Kohyama: Simul. Mater. Sci. Eng. **4**, 397 (1996).
8. M. Kohyama, R. Yamamoto: Phys. Rev. B **50**, 8502 (1994).
9. M. Kohyama, S. Kose, M. Kinoshita, R. Yamamoto: J. Phys. Condens. Matter **2**, 7809 (1990).
10. M. Kohyama, S. Kose, R. Yamamoto: J. Phys. Condens. Matter **3**, 7555 (1991).
11. F.S. Ohuchi, M. Kohyama: J. Am. Ceram. Soc. **74**, 1163 (1991).
12. M.W. Finnis: J. Phys. Condens. Matter **8**, 5811 (1996).
13. J.R. Smith, T. Hong, D.J. Srolovitz: Phys. Rev. Lett. **72**, 4021 (1994).
14. N. Troullier, J.L. Martins: Phys. Rev. B **43**, 1993 (1991).
15. J.P. Perdew, A. Zunger: Phys. Rev. B **23**, 5048 (1981).
16. K. Hiraga: Sci. Rep. Res. Inst. Tohoku Univ. A **32**, 1 (1984).
17. V.M. Bermudez: J. Appl. Phys. **63**, 4951 (1988).
18. W.R.L. Lambrecht, B. Segall, O.K. Andersen: Phys. Rev. B **41**, 2813 (1990).
19. A.P. Sutton, R.W. Balluffi: *Interfaces in Crystalline Materials*, (Oxford University Press, Oxford, 1995) Chap. 11.

Ab initio Tensile Testing Simulation of Al, AlN and Al/AlN Composite

Shigenobu Ogata and Hiroshi Kitagawa

Micro-dynamics Laboratory, Department of Adaptive Machine Systems, Osaka University, 2-1,Yamada-oka, Suita, Osaka 565-0871, Japan

Summary. The ideal strength and mechanical behavior of aluminum, aluminum nitride, and their composite model are investigated at zero temperature using the *ab initio* norm conserving pseudopotential method based on a local density approximation. The total energy and stress of supercells which contain these atomistic models are calculated under uniaxial tensile strain. Various estimated physical properties of aluminum and aluminum nitride at the equilibrium state are in good agreement with experiment. The tensile test of the composite model shows failure in the aluminum matrix. The ideal strength of the composite is also estimated.

Key Words. Tensile testing, density functional theory, *ab initio* molecular dynamics, ceramics metal interface, aluminum nitride, aluminum

1. Introduction

An aluminum based composite material including nano AlN particles has low thermal expansion, high thermal conductivity, high tensile strength and high ductility [1]. Also, it is known that these excellent properties are better than Al-matrix sapphire-particle or Al-matrix silicon carbide-particle composites [1,2]. For these reasons, it is expected to find use as a high performance heat sink. Lai and Chung [2] proposed that these properties are due to the strength of the atomic bonds on the interface between the Al-matrix and the AlN-particle. Therefore, details of the interface bonding properties should be investigated.

Recently, a number of continuum mechanics calculations have been performed to predict the mechanical behavior of composite materials. If these calculations are to be useful in predicting composite behavior, it is necessary to develop reliable models of mechanical response of interfaces in the composite materials. An atomistic simulation is a useful method to establish the models of specific interfaces because the macroscopic behavior is in principle determined by atomic level phenomena such as crack propagation via breaking of interatomic bonds. However, the reliability of these atomic simulations is strongly dependent on the model of the atomic interactions. The empirical and semi-empirical interactions have been successfully applied to calculations of metallic, ionic and covalent systems. However, if we simulate an interface between these systems, these theories have needed some adjustment of parameters or reformulation. On the other hand, *ab initio* simulations based on local density and pseudopotential approximations, which only require the atomic number and the atomic position, overcome the difficulty in estimating the atomic interactions between different kinds of atoms, and is known to give reasonable results for complex systems such as ceramics/metal interfaces [3,4].

The tensile test is the most efficient and usual approach to estimate the strength of materials. Recently, several tensile testing simulations using the *ab initio* calculation technique have been made. A simulation for TiC was done by Price *et al.* [5] using a full-potential-muffin-tin-orbital method. Deyirmenjian *et al.* [6] tested defective aluminum using a pseudopotential total energy calculation based on local density functional theory. However, there has been very little work on interfaces.

In the present work, the ideal strength and mechanical behavior of Al, AlN, and the Al/AlN interface at zero temperature under uniaxial tensile strain are investigated using the *ab initio* scheme. The stress–strain curves of these systems are also estimated.

2. Method

We used a plane wave representation for solution of the Kohn–Sham equations [7] which are derived on the basis of the density functional theory [8] and should be solved self-consistently. The local density approximation is adopted to describe the exchange-correlation energy as a functional of the electron density which was calculated by Ceperley and Alder [9] and parameterized by Perdew and Zunger [10]. The pseudopotential approximation is also used for representation of the electron-ion interactions in order to reduce a necessary number of plane waves. The pseudopotentials are generated by the scheme of Troullier and Martins [11] and represented in the form of Kleinman and Bylander [12]. The Brillouin zone (BZ) k integration is performed by the special points method of Monkhorst and Pack [13]. The occupation of states is determined by the Fermi distribution function. The Ewald summation method [14] is used to calculate the Coulomb contribution. The stress tensor is calculated using the explicit symmetric formulas proposed by Lee *et al.* [15] which is based on the Nielsen–Martin scheme [16].

$$\sigma_{\alpha\beta} = \frac{1}{\Omega(2-\delta_{\alpha\beta})}\frac{\partial E_{tot}}{\partial \varepsilon_{\alpha\beta}} \tag{1}$$

where the total energy per unit cell volume Ω is

$$E_{tot} = \left\langle \sum_i \frac{p_i^2}{2m_i} + E_{pot} \right\rangle . \tag{2}$$

Here, i labels atoms and angular brackets indicate the expectation value. E_{pot} denotes potential energy per unit cell. The wave vector $\boldsymbol{K} = \boldsymbol{k} + \boldsymbol{G}$ is transformed into

$$\frac{\partial K_\gamma}{\partial \varepsilon_{\alpha\beta}} = -\frac{1}{2}(\delta_{\alpha\gamma}K_\beta + \delta_{\beta\gamma}K_\alpha)(2-\delta_{\alpha\beta}) \tag{3}$$

under a symmetric strain $\varepsilon_{\alpha\beta}$. The conjugate gradient algorithm proposed by Bylander *et al.* [17] is employed to achieve a self-consistent electronic ground state. It is known that this scheme is the most efficient for ceramics/metal interfaces [3]. The motion of the atomic nuclei is calculated by integration of the classical equation of motion subject to the Hellmann–Feynman force. Relaxation of the atomic configuration is done by a simulated annealing technique.

The calculation presented in this paper is performed using our developed *ab initio* total energy calculation package on an NEC SX-4.

3. Model and Computational Conditions

The calculations for pure Al and AlN are performed using a 6 atom periodically repeated supercell with a plane wave cutoff energy of 20 Ry and an 8 atom supercell with the cutoff energy of 64 Ry, respectively. The tensile axis is set to Al[111] and AlN[0001] directions which corresponds to the z-direction of these supercells. Brillouin zone sampling is accomplished by a 4×2×2 Monkhorst–Pack scheme for the Al and the AlN supercell. Calculated equilibrium lattice constant, bulk modulus, cohesive energy, elastic constants, and surface energy of Al(111) are in good agreement with experimental data and previous calculations as Table 3.1 shows. The bulk moduli are estimated by pressure–volume curves (Fig. 3.1), and the surface energy is calculated with a supercell which has 6 Al(111) slabs and 5 interlayer spacings.

For the Al/AlN interface calculation, we use an 18 atom periodically repeated supercell involving a 4 layer Al(111) and N-terminated 5 layer AlN(0001) (Fig. 3.2). This model contains two inequivalent interfaces. It is

Table 3.1. Calculated equilibrium lattice constants, bulk modulus, cohesive energy, elastic constants and surface energy compared with experiment and other calculation

Al	Present calc.	Other calc.	Expt.
Surface energy (111) (J m^{-2})	1.10	0.97 [19]	—
Lattice constant (Å)	4.01	3.98 [6]	4.02 [21]
Cohesive energy (eV $atom^{-1}$)	3.12	—	3.39 [20]
Bulk modulus (GPa)	79.0	81 [6]	79.4 [18]
AlN	**Present calc.**	**Other calc.**	**Expt.**
Lattice constant a (Å)	3.070	3.144 [23]	3.110 [22]
Lattice constant c (Å)	4.920	5.046 [23]	4.980 [22]
Lattice constant u	0.382	0.381 [23]	—
Cohesive energy (eV $pair^{-1}$)	10.2	12.74 [25]	11.6 [24]
Bulk modulus (GPa)	207	195 [23]	208 [22]
Elastic constant C_{33} (GPa)	415	—	395 [27]

known that the matching of these crystallographic orientations is the most stable [28]. The Al(111) layers are expanded by 9.2% in the X−Y plane so as to be lattice matched to the AlN layers. The initial shape of the supercell is relaxed so that the all elements of the stress tensor of the supercell become sufficiently small ($\sigma_{\alpha\beta} < 0.1$ GPa). The plane wave cutoff energy of 50 Ry and 3×3×1 Monkhorst−Pack Brillouin zone sampling are used. The tensile axis is set to the normal direction to these interface.

The uniaxial tensile test is simulated by increasing the uniaxial tensile strains along the z-direction from $\varepsilon = -0.08$ to 0.27 for the Al, −0.08 to 0.33 for the AlN and −0.06 to 0.13 for the interface model in small increments. The other supercell dimensions are held fixed. The atomic configurations of the composite model are relaxed at each strain level. The relaxed atomic configuration are used as the initial configuration for the next stain level.

4. Results and Discussion

Fig. 4.1 shows the stress−strain curves of the Al, the AlN, and the Al/AlN composite model. The stresses σ_3 are measured on the surface of the supercell normal to the z-direction. The elastic constant C_{33} of AlN can be calculated from the slope of the curve at zero strain. The estimated value is in good agreement with experiment as is shown in Table 3.1. Beyond strains of ε =0.21 for Al, 0.32 for AlN, the calculated stress decreases gradually. This is due to the atomic configurations being unstable under this condition. An ideal strength of materials can be measured by the maximum stress value of

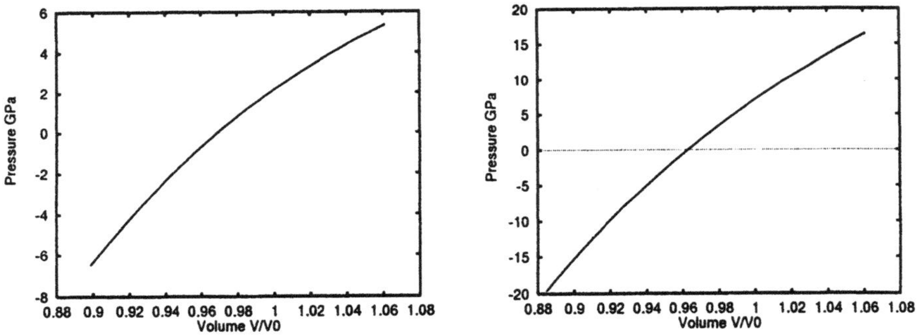

(a) Al (b) AlN

Fig. 3.1. Pressure–volume curves of Al and AlN. The volumes V are normalized against the experimental volume V_0.

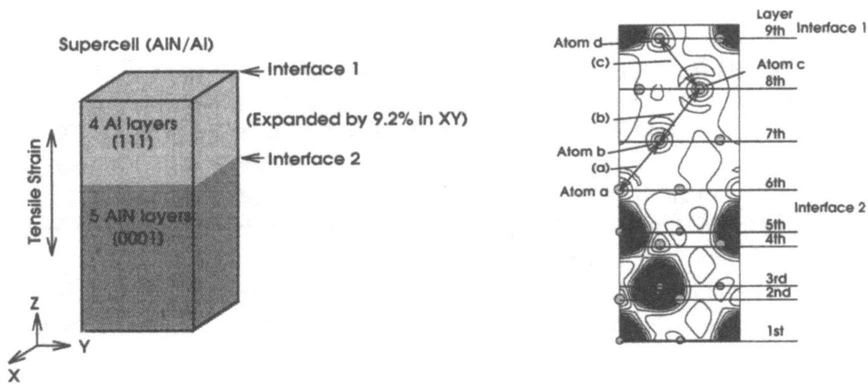

Fig. 3.2. Schematic illustration of Al(111)/AlN(0001) interface model.

the stress–strain curve. The maximum stress of Al and AlN are ~11 GPa at a strain $\varepsilon = 0.21$ and ~50 GPa at a strain $\varepsilon = 0.32$, respectively. On the other hand, the stress–strain curve of the Al/AlN composite model becomes nearly flat at 0.13 strain. The maximum stress is ~18 GPa. This value is 1.6 times larger than that of Al. This result is consistent with the fact that the Al matrix is reinforced by the AlN particle. The atomic configuration of the Al/AlN composite model is fully relaxed using the simulated annealing technique. The stress of relaxed supercell at the 0.07 and 0.13 strain are 2 GPa and 7 GPa lower than the stress of the uniformly strained and unrelaxed supercell, respectively. There is no relaxation of the stress up to ~0.05 strain and beyond that value the relaxation increases abruptly.

Fig. 4.2 illustrates the relaxed atomic configurations and contours of the valence electron density at each strain level ($\varepsilon = 0$, 0.07, and 0.13). The valence electron density distribution in Al and AlN layers are similar to their bulk states except for the interface. Moreover, any large disturbance of the atomic configuration cannot be found even at the interface. This observation contrasts strongly with the calculation results of SiC/Al interface [3], in which a large atomic movement and reaction occur at the interface. In AlN, the valence electrons are localized at N sites. Therefore, the bonding is more ionic than covalent. However, the tetrahedral coordination suggests that the covalency (sp^3 bonding) is also present. This fact is consistent with the result of other calculations for pure AlN [24]. The interfacial Al atoms of the Al(111) are captured by the interfacial N atoms of the AlN(0001), and sp^3 bonding occurs between these atoms. At 7% strain, the Al matrix is strained almost uniformly, while at 13% strain, only interlayer distances between the Al layers near interface 1 are increased. This fact suggests that the interfacial bond between Al and N atom is very strong and a failure of the composite occurs in the Al matrix in the neighborhood of the interface.

Table 4.1 shows changes of interlayer distances of the relaxed supercell at 0.13 strain. This result also shows the failure occurs in the Al interlayer, in the neighborhood of interface 1. We also calculate the case of an artificial failure occuring on the Al interlayer in the neighborhood of interface 2 so as to move the Al layers to the upper AlN layers. The other computational conditions are set to the same as for the normal calculation. A calculated total energy per supercell of the latter system is 0.2 eV higher than the normally calculated system. Hence, we find that the Al bonding near interface 1 is a little weaker than that near interface 2.

5. Conclusion

The uniaxial tensile simulations of Al, AlN, and Al/AlN interface model were conducted using the *ab initio* molecular dynamics method based on the

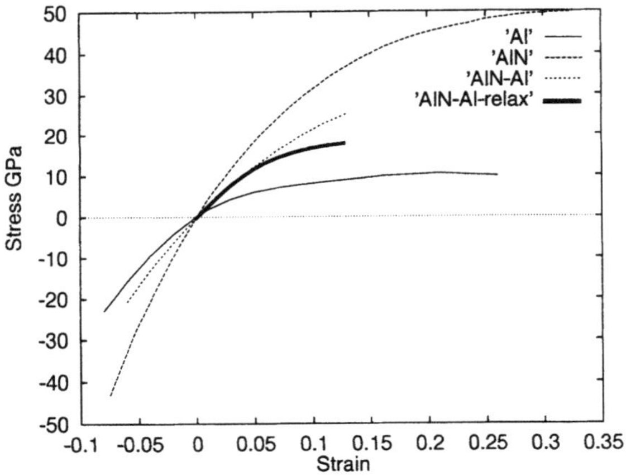

Fig. 4.1. Stress–strain curves of Al, AlN, and Al/AlN interface model under uniaxial tensile strain. The thin solid line and the broken line represent the stress of uniformly strained Al and AlN supercell, respectively. The heavy solid line and the dotted line represent the stress of relaxed and uniformly strained (unrelaxed) Al/AlN interface model, respectively. The stress are measured on the surface of the supercell normal to z-direction.

Table 4.1. Changes of the interlayer distance at $\varepsilon = 0.13$

Layer A-B	$\Delta d_{AB}/d^0_{AB}$
1-2	−0.071
2-3	−0.042
3-4	−0.072
4-5	−0.024
5-6	−0.071
6-7	−0.014
7-8	−0.020
8-9	0.216
9-1	0.034

d^0_{AB} is initial interlayer distance between A and B at $\varepsilon = 0$

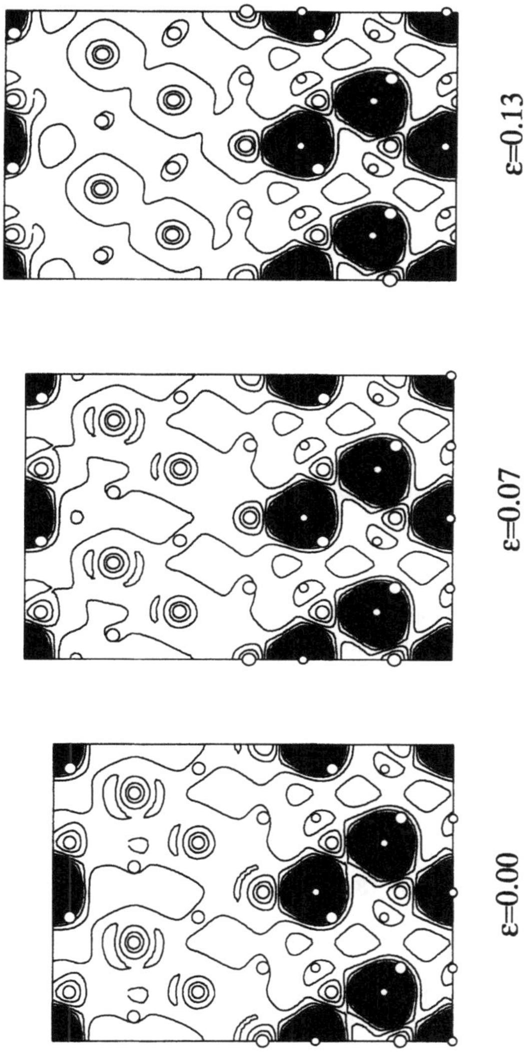

Fig. 4.2. Relaxed atomic configurations and contours of the valence electron density at each strain level (ε=0, 0.07, and 0.13). Large and small circles represent aluminum and nitrogen atoms, respectively. The interval of contours is 1×10^{-2} atomic unit.

density functional theory. The main results obtained are shown below.

- The ideal strength of Al and AlN were estimated.
- The ideal strength of Al(111)/AlN(0001) interface model was estimated.
- The failure of the Al/AlN composite model occurs at the Al interlayer in the neighborhood of the Al/AlN interface.

We plan in the near future to permit changes of the supercell shape in the relaxation processes using the Parrinello-Rahman scheme.

References

1. S.W. Lai, D.D.L. Chung: *6th International SAMPE Electronics Conference 1992*, 459.
2. S.W. Lai, D.D.L. Chung: *Electronic Packaging Materials Science VII Symposium 1994*, 207.
3. M. Kohyama: Model. Simul. Mater. Sci. Eng. **4**, 397 (1996).
4. S.Ogata, H.Kitagawa: J. Jpn. Inst. Metals **66**, 1079 (1996).
5. D.L. Price, B.R. Cooper, J.M. Wills: Phys. Rev. B **46**, 11368 (1992).
6. V.B. Deyirmenjian, V. Heine, M.C. Payne, V. Milman, R.M. Lynden-Bell, M.W. Finnis: Phys. Rev. B **52**, 15191 (1995).
7. W. Kohn, L.J. Sham: Phys. Rev. A**140**, 1133 (1965).
8. P. Hohenberg, W. Kohn: Phys. Rev. B**136**, 864 (1964).
9. D.M. Ceperley, B.J. Alder: Phys. Rev. Lett. **45**, 566 (1980).
10. J.P. Perdew, A. Zunger: Phys. Rev. B **23**, 5048 (1981).
11. N. Troullier, J.L. Martins: Phys. Rev. B **43**, 1993 (1991).
12. L. Kleinman, D.M. Bylander: Phys. Rev. Lett. **48**, 1425 (1982).
13. H.J. Monkhorst, J.D. Pack: Phys. Rev. B **13**, 5188 (1976).
14. P.P. Ewald: Ann. Phys. **64**, 253 (1921).
15. O.H. Nielsen, R.M. Martin: Phys. Rev. B **32**, 3792 (1985).
16. I.H. Lee, S.G. Lee, K.J. Chang: Phys. Rev. B **51**, 14697 (1995).
17. D.M. Bylander, L. Kleinman, S. Lee: Phys. Rev. B **42**, 1394 (1990).
18. G.N. Kamm, G.A. Alers: J. Appl. Phys **35**, 327 (1964).
19. N. Chetty, K. Stokbro, K.W. Jacobsen, J.K. Nørskov: Phys. Rev. B **46**, 3798 (1992).
20. C, Kittel: *Introduction to Solid State Physics 6.*, (John Wiley and Sons, Inc., New York, 1986).
21. P.K. Lam, M.L. Cohen: Phys. Rev. B **24**, 4224 (1981).
22. K.M. Taylor, C. Lenie: J. Electrochem. Soc., **107**, 308 (1960).
23. K. Miwa, A. Fukumoto: Phys. Rev. B **48**, 7897 (1993).
24. W.Y. Ching, B.N.Harmon: Phys. Rev. B **34**, 5305 (1986).
25. R.L. Walter, B. Segall: Phys. Rev. B **43**, 7070 (1991).
26. M. Ueno, A. Onodera, O. Shimomura, K. Takemura: Phys. Rev. B **45**, 10123 (1992).
27. M.E. Sherwin, T.J. Drummond: J. Appl. Phys. **69**, 8423 (1991).
28. A. Inoue, B.G. Kim, K. Nosaki, T. Yamaguchi, T. Masumoto: J. Appl. Phys. **71**, 4025 (1992).

Electronic Structure Approach to Hydrogen Embrittlement in fcc Transition Metals

S. Shimamura and S. Tanimori

Department of Applied Science, Faculty of Engineering, Yamaguchi University, Ube 755, Japan

Summary. The electronic origin of hydrogen embrittlement in fcc transition metals is studied by comparing electronic structure variations with lattice displacement in Pd and Ag. The electronic structure variations with cleaving and slip on the (111) plane have been calculated using the tight binding method that reproduces *ab initio* band structure calculations. The computed results show that lattice displacement increases the electron energy of Pd relative to that of Ag. In other w ords, electrons in Pd have a higher resistance to lattice displacement than those in Ag. This is caused by a difference in contributions of d and s-p electrons to the energy variation between Pd and Ag. Based on the computed results, the electronic origin of hydrogen embrittlement in Pd is discussed.

Key Words. Hydrogen embrittlement, transition metals, Pd, Ag, electronic structure

1. Introduction

Hydrogen embrittlement in metals has been studied for many years since the role of dissolved hydrogen in deformation and fracture was recognized for steel materials. A number of investigations on this subject have been reported and reviewed in the literature to date. However, the origin of hydrogen embrittlement is not clear yet from a microscopic point of view.

In this paper we report the results of electronic structure calculations that have been carried out in order to understand the origin of hydrogen embrittlement in fcc transition metals. In the next section we briefly review the mechanism of hydrogen embrittlement in metals and point out an unsolved problem from an electronic point of view. In Sect. 3 we develop an electronic approach to the problem. In this approach the electronic structure variation with lattice displacement has been calculated for a fcc transition metal, Pd and a noble metal, Ag. The computed results are shown in Sect. 4. In Sect. 5 we discuss the electronic origin of hydrogen embrittlement in Pd on the basis of the computed results.

2. Unsolved Problem in Hydrogen Embrittlement

There appear to be two convincing mechanisms of hydrogen embrittlement in metals: the decohesion mechanism and the hydrogen-enhanced local plasticity (HELP) mechanism. The decohesion mechanism was discussed by Oriani and Josephic [1], who showed that the critical stress-intensity factor under mode I loading in AISI 4340 steel decreases with increasing hydrogen gas pressure. In this mechanism, the presence of hydrogen decreases the interatomic bond strength in metals and facilitates cleavage fracture. The HELP mechanism was discussed by Lynch [2], who carried out bending tests and electron microscopy observation for various kinds of metals and showed that the presence of hydrogen induces localized plastic flow at crack tips. In this mechanism the presence of hydrogen enhances plastic deformation at crack tips and this locally enhanced ductility facilitates crack propagation resulting in apparent brittle fracture.

In recent years, several theoretical or computational studies have been reported on hydrogen embrittlement. Adachi and Imoto [3] carried out electronic structure calculations for small clusters of iron-series transition metals containing an interstitial hydrogen atom by using the discrete variational Xα (DV-Xα) method. They suggested that the electron transfer from a hydrogen 1s orbital to the metal antibonding level weakens the metal bond in later-group transition metals. Daw and Baskes [4] developed the embedded-atom method (EAM) to investigate hydrogen embrittlement. They showed that the presence of hydrogen reduces the fracture stress in Ni. Zhong *et al.* [5] performed molecular dynamics simulations of hydrogen embrittlement for 500 Pd atoms. They found that the presence of hydrogen has a similar effect on the critical tensile stress as a temperature increase. Their simulations suggested that the apparent hydrogen embrittlement results from a local enhancement of ductility in hydrogen-saturated regions of the metal, supporting the HELP mechanism.

Thus, recent theoretical or computational studies gradually reveal atomic and electronic processes in hydrogen embrittlement in metals. However, we here point out a fundamental problem in hydrogen embrittlement in metals. Even if dissolved hydrogen induces cleavage or slip in metals, what is the electronic origin of hydrogen-induced cleavage or slip? Although several theoretical or computational studies show that the presence of hydrogen reduces the fracture stress in metals or enhances ductility in metals, how can we understand this in terms of electronic states? This is an unsolved problem in hydrogen embrittlement in metals from a microscopic point of view. Knowledge of electronic processes will be required to solve this problem.

3. Electronic Approach

As a first step in the study of the electronic origin of hydrogen embrittlement in metals, we consider electronic states in displaced metal lattices in the absence of hydrogen. A transition metal, Pd and a noble metal, Ag with the fcc crystal structure have been chosen to compare the roles of d and s-p electrons in lattice displacement. We calculate electronic structure variations with cleaving and slip on the (111) plane. As shown in Fig. 1, the (111) planes in the fcc lattice are periodically cleaved or slipped. In the cleaved lattice the (111) planes are cleaved at intervals of six layers, and in the slipped lattice they are slipped by turns at intervals of three layers; the slip direction is $[\bar{1}10]$ [6]. A unit cell contains six layers in both lattices.

The calculations have been carried out using the tight binding method (the linear combination of atomic orbitals approximation) [7]; 4d, 5s and 5p atomic orbitals for Pd and Ag have been used assuming that these orbitals are orthogonal. The Hamiltonian matrix elements including the first and the second neighbor interactions have been taken into consideration in the two-center approximation [8, 9]. The matrix elements in the undisplaced fcc lattice have been evaluated by using the Slater and Koster parameters [8, 9] given by Papaconstantopoulos [10]; this method reproduces first-principles APW band structure calculations for Pd and Ag [10]. The matrix elements in the displaced lattices have been evaluated by assuming that the Slater and Koster parameters decrease exponentially with the interatomic distance. The density of states has been calculated using the method by Rath and Freeman [11].

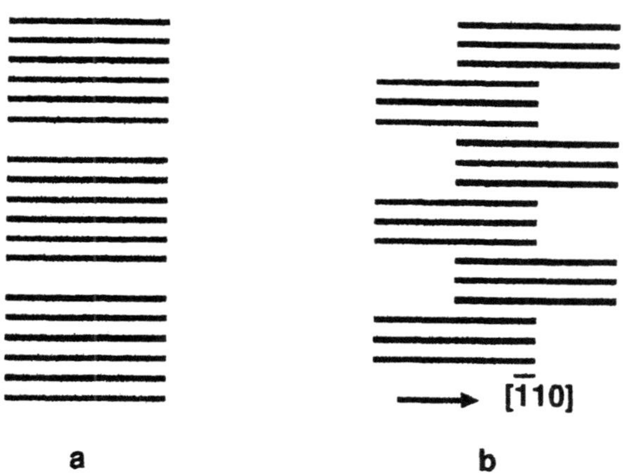

Fig. 1. Schematic representations of periodically cleaved and slipped (111) planes: (**a**) for the cleaved lattice and (**b**) for the slipped lattice.

4. Computed Results

We here show the computed results of the density of states (DOS) per unit cell and the electron energy in the cleaved and slipped lattices. We focus, in particular, on the difference between Pd and Ag.

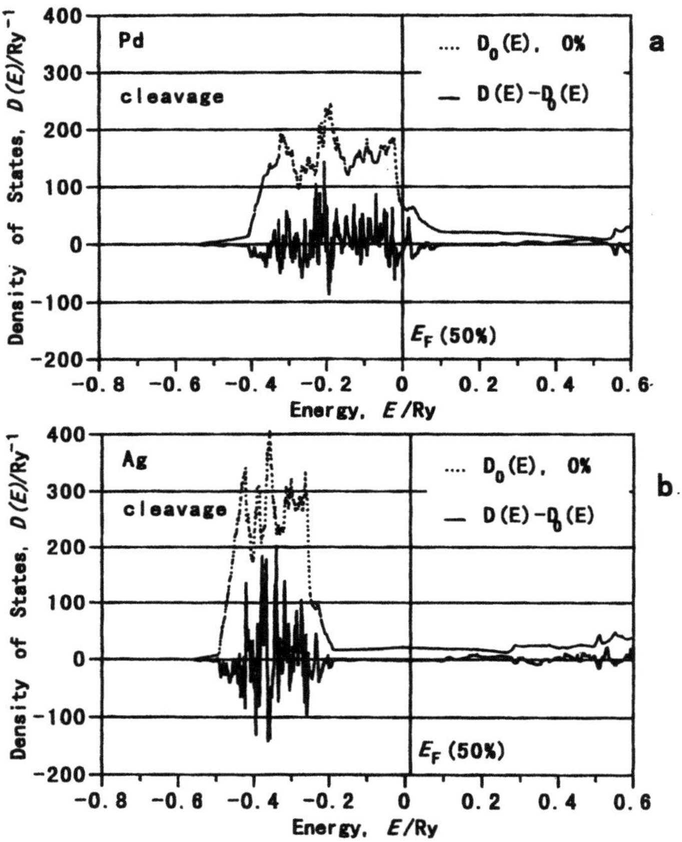

Fig. 2. Change in the density of states with cleaving: (**a**) for Pd and (**b**) for Ag; $D_0(E)$ for the uncleaved lattice (dotted line); $D(E)$ for the cleaved lattice with $\delta = 50\%$; the difference $D(E) - D_0(E)$ (solid line).

In the cleaved lattice, the degree of cleaving is represented by the deformation ratio, δ, which is defined as $(s - s_0)/s_0 \times 100$ %, where s is the spacing between cleaved (111) planes and s_0 is the spacing between (111) planes in the uncleaved lattice. Figure 2 shows the change in the DOS with cleaving for Pd and Ag: $D_0(E)$ (dotted line) is the DOS for the uncleaved fcc lattice, $D(E)$ is for the cleaved lattice with $\delta = 50\%$, and $D(E) - D_0(E)$ (solid line) is the difference. The vertical line denotes the position of the Fermi level, E_F for the cleaved lattice with $\delta = 50\%$. As shown by the difference, the DOS decreases in the low and high energy regions in the d bands and increases in the middle regions with cleaving for both metals. In other words, the d band width decreases due to cleaving. This is attributed to the fact that a decrease in the interatomic transfer of electrons across the cleavage plane causes a reduction in band dispersion in the direction perpendicular to the cleavage plane.

There is a quantitative difference in the change in the DOS with cleaving between Pd and Ag. The narrowing of the d band width is more conspicuous in Pd than in Ag. This is due to the fact that the d bands in Pd are located near E_F.

The narrowing of the band width in the cleaved lattice causes an increase of the electron energy. Figure 3 shows the deformation ratio dependence of the electron energy for Pd and Ag. The difference between Pd and Ag is manifested by the contributions of d and s-p electrons to the energy variation with cleaving. Because the existence of unoccupied states in the d bands for Pd pushes up the electron energy due to cleaving, the energy increase for Pd is dominated by d electrons; the energy of s electrons rather decreases with cleaving. On the contrary, the energy increase for Ag is dominated by s-p electrons because the d bands are fully occupied. The total electron energy (of 4d, 5s and 5p electrons) increases with cleaving for both metals.

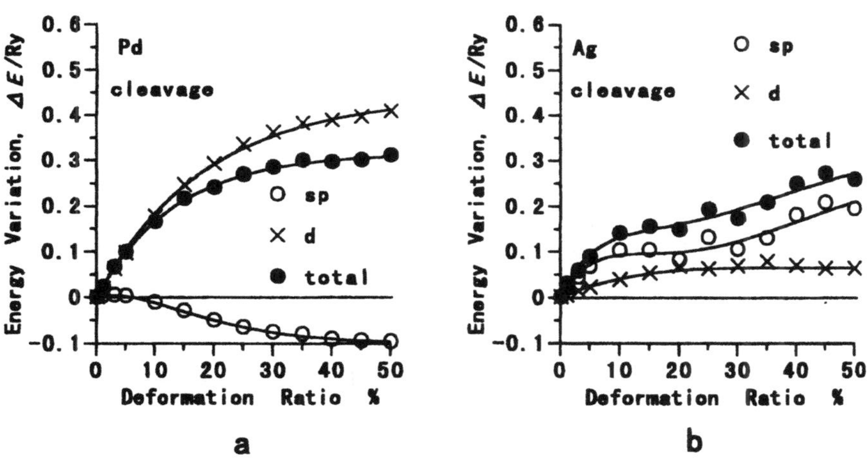

Fig. 3. Deformation ratio dependence of the electron energy in the cleaved lattice: (**a**) for Pd and (**b**) for Ag. The contributions of d and s-p electrons to the energy variation are also shown.

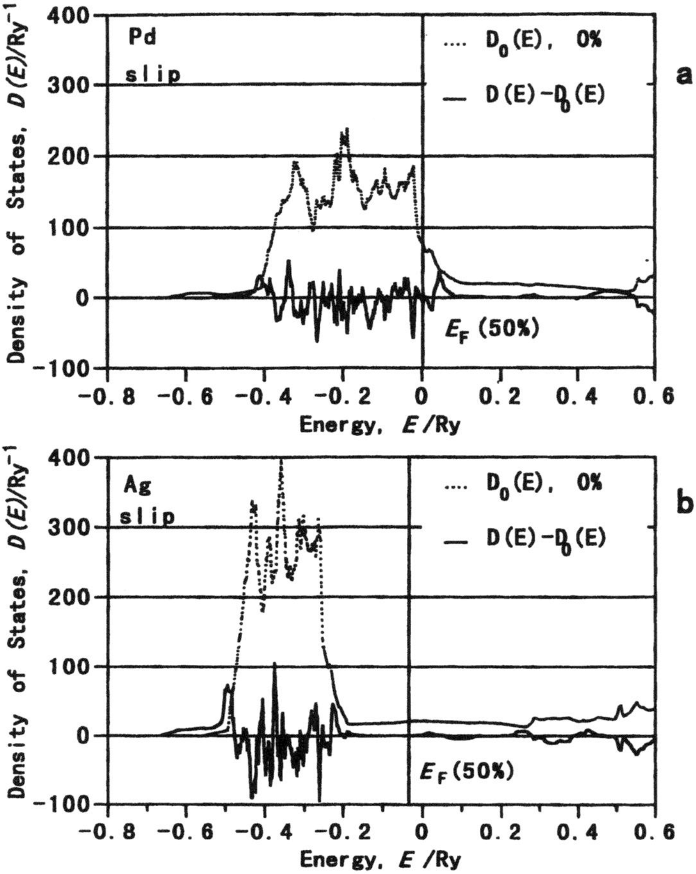

Fig. 4. Change in the density of states with slip: (**a**) for Pd and (**b**) for Ag; $D_0(E)$ for the unslipped lattice (dotted line); $D(E)$ for the slipped lattice with $\delta = 50\%$; the difference $D(E) - D_0(E)$ (solid line).

In the slipped lattice, the degree of slip is represented by the deformation ratio, δ, which is defined as $(\Delta d/d_0)\times 100\ \%$, where Δd is the slip distance and d_0 is the interatomic distance in the slip direction in the unslipped lattice. So $\delta = 100\%$ corresponds to slip of one interatomic distance in the slip direction, resulting in the unslipped lattice. Figure 4 shows the change in the DOS with slip for Pd and Ag: $D_0(E)$ (dotted line) is the DOS for the unslipped fcc lattice, $D(E)$ is for the slipped lattice with $\delta = 50\%$, and $D(E) - D_0(E)$ (solid line) is the difference. As shown by the difference, the DOS increases in the low and high energy regions in the d bands and decreases in the middle regions with slip for both metals. In other words, the d band width broadens due to slip in contrast with cleaving. This is attributed to the fact that the nearest-neighbor interatomic transfer of electrons near the slip plane increases with slip.

There is a quantitative difference in the change in the DOS with slip between Pd and Ag. The broadening of the d band width is more conspicuous in Pd than in Ag because the d bands in Pd are located near E_F.

The broadening of the band width in the slipped lattice causes a decrease of the electron energy. Figure 5 show the deformation ratio dependence of the electron energy for Pd and Ag. The difference between Pd and Ag is also manifested by the contributions of d and s-p electrons to the energy variation with slip. Although both d and s-p electrons contribute to the energy decrease, the predominant contribution is contrastive between Pd and Ag. Because the d bands are partially occupied for Pd and fully occupied for Ag, the energy decrease is dominated by d electrons for Pd and by s-p electrons for Ag.

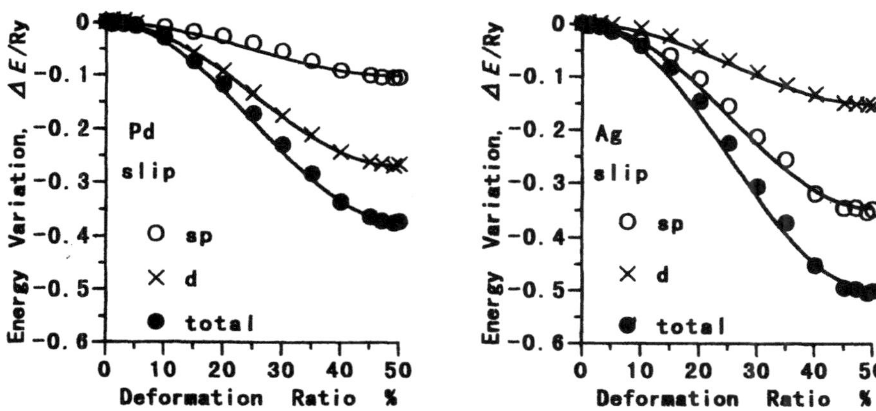

a b

Fig. 5. Deformation ratio dependence of the electron energy in the slipped lattice: (**a**) for Pd and (**b**) for Ag. The contributions of d and s-p electrons to the energy variation are also shown.

5. Electronic Origin of Hydrogen Embrittlement

We here discuss the electronic origin of hydrogen embrittlement in Pd by comparing the electron energy variations with lattice displacement for Pd and Ag. Figure 6 shows the variation of the total electron energy (of 4d, 5s and 5p electrons) versus the deformation ratio in the cleaved and slipped lattices for Pd and Ag. The electron energy increases with cleaving for both metals. However the energy increase for Pd is larger than that for Ag. In other words, electrons in Pd have a higher resistance to cleaving than those in Ag. This is due to the d electron contribution to the energy variation, as seen in Fig. 3. On the other hand, the electron energy decreases with slip for both metals in contrast to cleaving. The energy decrease for Pd is smaller than that for Ag. The larger energy decrease for Ag is due to the s electron contribution, as seen in Fig. 5. This also implies that electrons in Pd have a higher resistance to slip as well as cleaving than those in Ag.

It should be noted that the energy shown in Fig. 6 is not the total lattice energy but the electron energy; the energy decrease with slip does not represent the instability to slip. Figure 6 demonstrates that Pd is more resistant to lattice displacement than Ag, from an electronic point of view. This is caused by the fact that the electronic response to lattice displacement is dominated by d electrons for Pd and by s-p electrons for Ag, as shown in the preceding section.

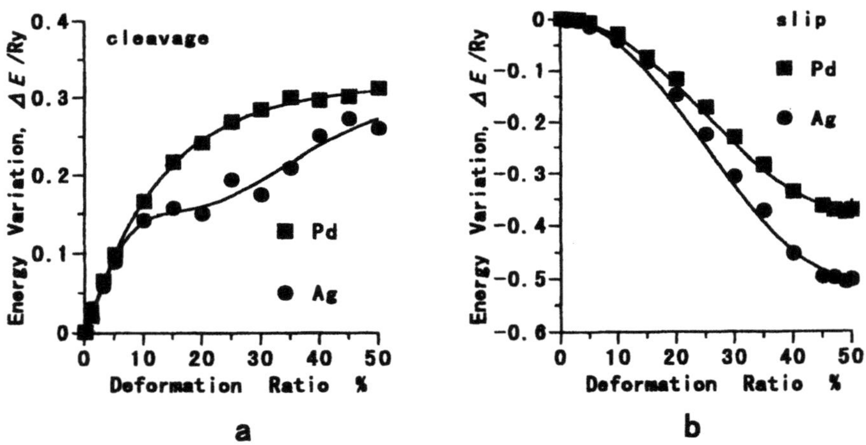

Fig. 6. Electron energy variation with lattice displacement for Pd and Ag: (**a**) for cleaving and (**b**) for slip.

A number of experimental investigations show that dissolved hydrogen atoms gather near crack tips in metals under stress; regions near crack tips are almost hydrogen-saturated. The electronic structure near crack tips in hydrogen-loaded Pd is similar to that in PdH. Figure 7 shows the density of states for PdH calculated by Papaconstantopoulos [12]. An important difference in electronic structure between Pd and PdH is the movement of the Fermi level up to the s-p bands. Thus, the electronic structure of PdH is similar to that of Ag in the sense that the Fermi level is on the s-p bands. Therefore we can infer that the electronic response to lattice displacement near crack tips in hydrogen-loaded Pd is dominated by s-p electrons. This will cause a decrease of the electronic resistance to metal bond breaking near crack tips in hydrogen-loaded Pd. Thus, a local change in electronic structure is responsible for hydrogen-assisted cracking.

The electronic origin proposed here relies only on the electronic response to lattice displacement. This is justified by the fact that cohesive properties of transition and noble metals are dominated by valence d and s-p electrons, respectively. In fact, differences in the cohesive energy, elastic constants and the melting temperature between transition and noble metals can be understood by a difference in the position of the Fermi level relative to the d bands.

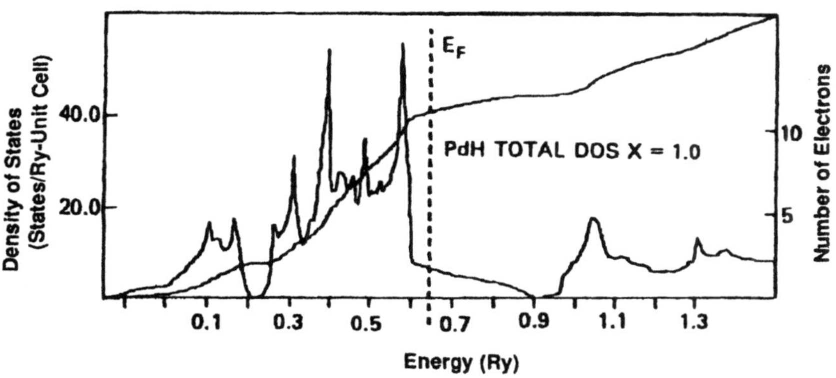

Fig. 7. Density of states in PdH (from [12]).

In the present approach, a simple tight binding method has been used to calculate electronic structures in cleaved and slipped lattices for Pd and Ag. First-principles calculations will be required to discuss detailed variations in electronic structure with lattice displacement. Also, calculations of electronic structure variations with lattice displacement for PdH are left for future study. The present study gives some insight into a microscopic understanding of hydrogen embrittlement in transition metals.

References

1. R. A. Oriani, P. H. Josephic: Acta Metall. **22**, 1065 (1974).
2. S. P. Lynch: Acta Metall. **36**, 2639 (1988).
3. H. Adachi, S. Imoto: J. Phys. Soc. Jpn. **46**, 1194 (1979).
4. M. S. Daw, M. I. Baskes: Phys. Rev. Lett. **50**, 1285 (1983).
5. W. Zhong, Y. Cai, D. Tománek: Nature **362**, 435 (1993).
6. C. Kittel: *Introduction to Solid State Physics*, 7th ed., (John Wiley & Sons, New York, 1996).
7. W. A. Harrison: *Electronic Structure and the Properties of Solids: The Physics of the Chemical Bond*, (W. H. Freeman and Company, San Francisco, 1980).
8. J. C. Slater, G. F. Koster: Phys. Rev. **94**, 1498 (1954).
9. R. R. Sharma: Phys. Rev. B **19**, 2813 (1979).
10. D. A. Papaconstantopoulos: *Handbook of the Band Structure of Elemental Solids*, (Plenum Press, New York, 1986).
11. J. Rath, A. J. Freeman: Phys. Rev. B **11**, 2109 (1975).
12. D. A. Papaconstantopoulos: *Proc. NATO Advance Study Institute on Metal Hydrides*, ed. G. Bambakidis, (Plenum Press, New York, 1981) p.215.

Catastrophic Transformation of Electron Stress and Electron Stiffness Parameter on Metal and Semiconductor

Shigeo Kotake, Hiroyuki Kimata, Takashi Aoki,
Yasuyuki Suzuki and Masafumi Senoo

Department of Mechanical Engineering, Mie University, 1515 Kamihama-cho,
Tsu city, Mie 514, Japan

Summary. In this study, we calculated the quantum stress of electrons, called electron stress (ES), in a metal and a semiconductor under uniaxial tensile and compressive deformation (-50% to 50%) by using the pseudopotential method. Since the positions of atoms are fixed with external forces, the ES of a stationary electron shows the internal stress among atoms. From the derivative of ES with strain, we defined a "quantum stiffness parameter (ESP)". We calculated ES and ESP in Al, K and Si during deformation. From the results of the calculation, we consider the following points: (i) Change of ES and ESP during deformation, (ii) ES and ESP demonstrate the difference between a metal and a semiconductor. During uniaxial tensile deformation, after showing a gradual decrease, ESP shows a sudden divergence from positive to negative values at a certain critical strain. Under compression, a merely gradual increase of ESP was observed. Strain at the critical transformation of ESP, termed "critical strain (ε_{cr})", shows a material dependence; 20% in Al, 30% in K, but only 2.5% in Si. The result of a low ε_{cr} in Si and high ε_{cr} values in metals corresponds to the general knowledge on mechanical properties of materials; metals are more ductile than semiconductors. The critical transitions of ES and ESP in solids could be explained from the electric properties in solids. These parameters might provide a key to understand ideal fractural properties in solid.

Key Words. Fracture mechanics, electron stress, electron stiffness parameter, pseudopotential method, critical strain, band structure

1. Introduction

Fracture is one of the universal properties of materials, which has been explained from an energy balance, such as Griffith's theory, rather than a force balance. Griffith's theorem introduced the basic concept of fracture toughness (K_{Ic}), which treats the fractural phenomenon as a catastrophic transition from elastic energy to surface energy. In practice, since elasticity and plasticity simultaneously exist during deformation, fractural phenomenon becomes rather complicated. Although the concept of K_{Ic} is quite important in the engineering field, it remains an empirical quantity.

Generally speaking, metals are ductile and semiconductors (or insulators) are brittle. Some metals show large plastic deformation during uniaxial tensile testing, whereas semiconductors show little plastic deformations. Since electron theory well explains the drastic change of physical properties between metals and semiconductors, such as electric conductivity and optical phenomena, certain relations between band structure and fractural behavior are implied. Since band structure is the major difference between metals and semiconductors, electron theory might be key to understand the difference in fractural strain. However, since electron theory can only treat small periodic systems, the calculations are confined to idealized generalities. The practical fractural phenomena, including dislocations and cavities, are beyond this method.

Recently, progress in analytical calculations and simulations on fracture has been made. Total energy in the electron theory [1, 2] showed ideal strength from instabilities of bulk modulus. Although ideal strength has been discussed since the early stages of electron theory, expected values only have been noted.

In recent years, rather than entire properties, the local distribution of physical quantities has been studied in molecular dynamics (MD) simulation [3]. To estimate atomistic distributions of mechanical properties, such as force and stiffness and mechanical properties in terms of electron theory, probability densities on physical stress of electron, called electron stress (ES), are proposed in this study.

The concept of stress in quantum mechanics was first introduced by Pauli [4] and other quantum physicists in the 1930s from the quantum force balance equation between the quantum virial force (F_V) [5] and the Hellmann-Feynman force (F_f) [6]. The parameter F_V, which is the divergence of ES, corresponds to the classical virial force in ideal gas theory. Under stationary condition, F_V is equal to F_f. Nakatsuji [7] was the first to note the importance of the probability density of physical quantities to chemical reactions. He developed a hypervirial theorem from the quantum force balance equation. Ghosh [8] has shown the application of the hypervirial theorem in an isolated atom. He showed the distribution of ES around a Kr atom. Nielsen and Martin [9] applied the hypervirial theorem to analyze mechanical properties in solids, especially stiffness. They obtained macroscopic stiffness from the first-principles calculation. After Folland [10] introduced the microscopic stress tensor, since most former works have been done on estimating expectation values in a whole system, there has been some controversy [11-13]. Evaluation of microscopic physical quantities, such as probability density, is a recently proposed new concept.

The authors have discussed the local distribution of ES and the phase transition under tensile and compressive deformation in fcc and bcc simple metals [14]. We proposed the electron stiffness parameter (ESP), that shows a bond-like image among atoms, from the derivative of ES with strain.

In this study we applied ES and ESP to analyze mechanical properties, especially lattice instabilities in solids. Under 3-dimensional periodic conditions, local distributions of ES and ESP have been calculated in Al,K and Si under uniaxial deformation. The crystal structure of Al, K and Si is fcc, bcc and diamond type structure, respectively. Under each strained condition, since atomic configuration is fixed with the stationary condition of the electrons, force is applied from electrons to atoms in the system, as shown in Fig. 1.1. From the force balance equation, the change of ES corresponds to the applied force between electrons and atoms under tensile and compressive conditions. According to the Hellmann-Feynman theorem, forces among atoms consist of not only electron-ion interactions, but also ion-ion interactions. Although ESP does not directly connect to actual stiffness of materials, since it expresses a local density of electron-atom interactions, it is one of the important parameters to estimate internal properties.

Since fracture at ideal strength, termed ideal fracture, coincides with a catastrophic transition in perfect crystals, certain signs, such as lattice instabilities, are expected on ES and ESP images. Then how can ideal fracture be estimated in terms of ES and ESP?

In this calculation, we are concerned with the following questions;

(1) How does the distribution of ES and ESP change during deformation?

(2) Do catastrophic phenomena exist in ES or ESP at a certain strain?

(3) Are there any differences between ES and ESP in metals and those in semiconductors?

(4) Does ideal fracture occur in specific bonds or not?

(5) What does the behavior of ESP correspond to, especially in terms of band structure?

Since ES and ESP can express spatial distribution of mechanical quantities of each electron, they should show new concepts for mechanical properties.

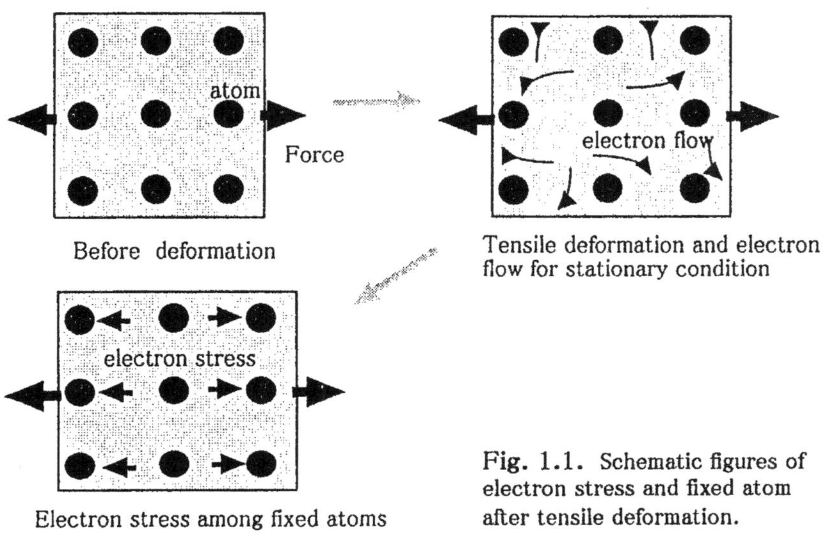

Fig. 1.1. Schematic figures of electron stress and fixed atom after tensile deformation.

2. Theorem

In this section we will show the derivation of ES and ESP from the Schrodinger equation. These equations are known as the "quantum force balance [8]".

2.1 Probability Density

In general, the probability density of a physical quantity is a complex value, whose real term is closely related to the distribution of physical quantities in classical mechanics, such as density of probability (ρ) and probability current density (ι). Since all quantum operators are Hermitian, whole integration of the real terms (expectation values) is equal to the law of classical physics; Eherenfest's theorem. On the other hand, whole integration of the imaginary terms tends to zero. Since ES and ESP are the real terms of probability density, they correspond to the distribution of stress and stiffness within a solid.

2.2 Continuum Equation and Probability Current Density

The term ρ, which is the square of the norm of the wave function (ϕ (k)), expresses the probability of existence, as shown in (2.1). The parameter ι is the real term of the probability density of momentum, as shown in (2.2). Under the polar coordinate, wave function is expressed with ρ and phase (S) in (2.3). Similarly, ι is expressed with ρ and S in (2.4). Here, $\hbar = 2m = e^2/2 = 4\pi\varepsilon_0 = 1$, as atomic unit (a.u.).

$$\rho = \phi^* \phi \tag{2.1}$$

$$\iota = i(\phi^* \partial \phi / \partial x - \phi \partial \phi^* / \partial x) \tag{2.2}$$

$$\phi = \rho \exp(i S) \tag{2.3}$$

$$\iota = \rho \nabla S \tag{2.4}$$

The continuum equation (2.5), which represents the divergence of ι which causes the evolution of ρ, can be obtained from the derivative of ρ with time. This equation shows temporal conservation and spatial continuum of ρ.

$$\partial \rho / \partial t + \nabla \cdot \iota = 0 \tag{2.5}$$

2.3 Force Balance Equation in Quantum Mechanics

From the derivative of ι with time, ι can also be expressed in the continuum equation (2.6). According to classical mechanics, the temporal derivative of momentum expresses force. The left side of (2.6) shows the probability density of the force on an electron, shown as the temporal evolution of ι. In the right side of (2.6), the first term represents the force from the gradient of potential, called the Feynman force, defined in (2.7). The second term represents the force from spatial derivative of momentum of the electron, called the virial force, defined in (2.8). The right side of (2.8) expresses the stress of the wave function, called the electron stress (σ_{ij}), as shown in (2.9). Therefore, the force balance between F_f and F_V affects the local evolution of ι; the local acceleration of electron. This equation shows the temporal conservation and the spatial continuum of ι. By using (2.5) and (2.6), the evolution equation of an electron [15], called the quantum fluid dynamics, has been proposed.

$$\partial \iota / \partial t = - F_f + F_V \tag{2.6}$$

$$F_f = \rho \nabla V \tag{2.7}$$

$$F_V = -\nabla \cdot \sigma_{ij} \tag{2.8}$$

$$\sigma_{ij} = 1/2 \left(\psi^* \nabla \nabla \psi + \psi \nabla \nabla \psi^* - \nabla \psi \nabla \psi^* - \nabla \psi^* \nabla \psi \right) \tag{2.9}$$

2.4 Electron Stress and Electron Stiffness Parameter

As shown in (2.9), ES is a second order tensor, which is the second spatial derivative of the wave function. Like the continuum equation of ρ, divergence of ES expresses F_V, that is analogous to the balance equation in continuum mechanics. In the system where wave functions are primarily determined, such as the MO method, ES is purely deduced from them. In continuum mechanics, a positive ES is defined as compressive stress and a negative ES is defined as tensile stress. Divergence of ES (F_V) works against the gradient of potential (F_f). This means ES is the origin of F_V which balances F_f under the stationary conditions of an electron, as shown in (2.6).

Analogous to stiffness in continuum theory, ESP, expressed as C_{ijkl}, is the derivative of ES with strain, shown in (2.10). In this study, C_{ijkl} is simply expressed as C_{mn}. In continuum mechanics, although stiffness of a solid is uniform under simple compression and tensile deformation, ESP shows an atomistic density image of stiffness. High ESP expresses strong spring constants in the atomistic field.

$$C_{ijkl} = \partial \sigma_{ij} / \partial \varepsilon_{kl} \tag{2.10}$$

3. Calculation

In this study we have calculated ES and ESP of Al, K and Si under uniaxial compressive and tensile deformation. Atomic configurations are assumed under the following conditions;

(1) Unit cell is under periodical conditions.

(2) In the unit cell, one lattice parameter is tensile and compressive deformed under -50% to 50% strain, while the other lattice parameters are fixed as constant.

(3) Ratios of atomic positions in unit cell are fixed.

(4) Pseudopotential parameters of atoms are constants under any deformed condition.

Like the phase transformation of martensite, the ratios of atomic positions are not always unchanged. In this study, force, which causes a phase transformation, might be observed as a change of ES or ESP, since atomic positions are fixed.

The wave functions of valence electrons under each atomic configuration are obtained from perturbation theory with the Heine-Abarenkov pseudopotential [1]. Concerning crystal symmetry, zero and first order wave functions of electron were expanded into plane waves within 3 k_f (k_f: Fermi wave number, q: reciprocal vector of lattice, a_q : perturbation expansion coefficient) in reciprocal space under stationary conditions (3.1). Table 3.1 shows the crystal structures, initial lattice constants (a_0), number of valence electrons (Z) and pseudopotential parameters (A, R_M). A and R_M are determined by Senoo et al.[1] to fit the stress-strain relations of materials under ultra high pressure. Since materials are highly deformed, these parameters are more suitable in this study.

$$\phi(k) = \Sigma\, a_q \,|\, k+q\rangle \tag{3.1}$$

Each axis of the unit cell is divided into 20 pieces to make 8000 local areas. From the secondary spatial derivative of the wave function in (2.9), the ES of each electron was obtained analytically in each local area. ES in each area is obtained by summation of ES of each electron in (3.2). Under tetragonal symmetry, four ES ($\sigma_{11}, \sigma_{12}, \sigma_{13}$ and σ_{33}) are independent. Hydrostatic pressure (P) of ES, as shown in (3.3), is also calculated,

$$\sigma_{ij}(x) = \Sigma\, \sigma_{kij}(x) \tag{3.2}$$

$$P = (\sigma_{11} + \sigma_{22} + \sigma_{33}) / 3 \tag{3.3}$$

Table 3.1. Crystal structure and parameters on Al, K and Si in pseudopotential calculations (a.u.).

	Al	K	Si
Crystal structure	fcc	bcc	diamond
a_0	7.65339	9.87382	10.26121
Z	3	1	4
A	0.88149	0.18658	1.2874
R_M	1.35	3.0054	1.1602

ESP(ε) under arbitrary strain (ε) was obtained from the finite difference between ES(ε) and ES(ε + 0.001) with strain. From (2.8), F_V is calculated from the divergence of ESP. F_f and ρ are also calculated from (2.7) and (2.1).

4. Results

4.1 Electron Stiffness Change under Uniaxial Tensile and Compressive Deformation of K, Al and Si

Figure 4.1-4.3 show (100) plane images of ESP (C_{11}) in K, Al and Si, respectively, under tensile strain. At the corner of the ESP image in K and at the center and the corner of ESP image in Al and Si, each atom is located. Maximum and minimum EP (σ_{11}) changes in K during uniaxial tensile deformation are shown in Fig. 4.4. During increasing strain from 0%, ESP shows a gradual decrease, but remains positive. On the other hand, after strain reaches a certain value, ESP catastrophically changes from a positive to a negative value. The strain of transformation is termed "critical strain (ε_{cr})".

Each ε_{cr} of Al, K and Si is 30%, 20% and 2.5%, as shown in Table 4.1. These catastrophic transformations strongly depend on the kind of materials, especially metals and semiconductors. Transformation of ESP has been observed in the whole area of a unit cell in all systems.

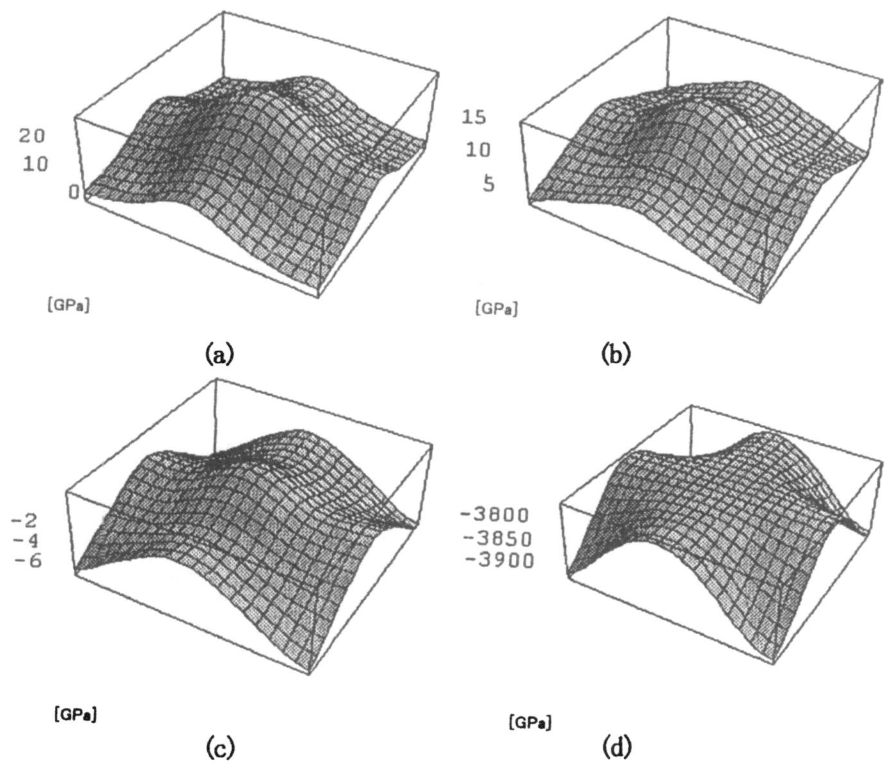

Fig. 4.1. C_{11} images of K (100) under (a) 0%, (b) 20%, (c) 30% and (d) 40% tensile strain.

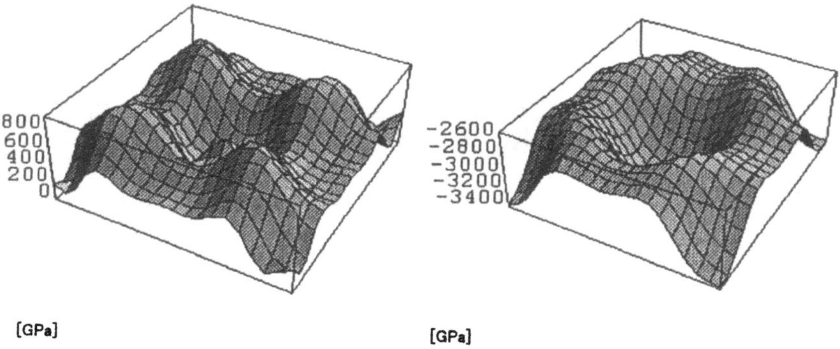

(a) (b)

Fig. 4.2. C_{11} images of Al (100) under (a) 0% and (b) 20% tensile strain.

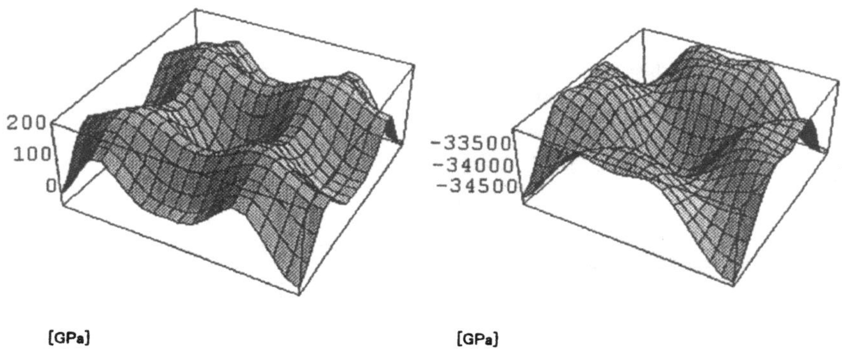

(a) (b)

Fig. 4.3. C_{11} images of Si(100) under (a) 0% and (b) 10% tensile strain.

Table 4.1. Critical strain of K, Al and Si

	K	Al	Si
ε_{cr} (%)	30	20	2.5

4.2 Virial Force, Feynman Force, Electron Stiffness Parameter and Density of Probability

F_V and F_f images of the K (100) plane under 0% strain are shown in Figure 4.5 a,b. Each atom is located at the corner. In F_f, lighter contrast expresses a higher absolute value of force, whereas in F_V, darker contrast does. As far as electrons are in stationary conditions, F_V is equivalent to F_f(2.6). Since the stationary wave function is approximately obtained from perturbation

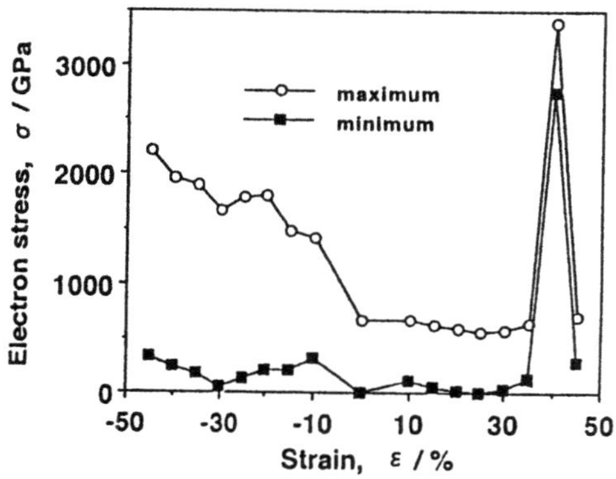

Fig. 4.4. Maximum and minimum of σ_{11} in deformed K.

theory, small differences exist between F_V and F_f. Since F_f originates from the gradient of the potential energy and F_V originates from the wave function itself, the meanings of each method are different. In this pseudopotential method, although the wave function is determined from the pseudopotential, material properties are calculated from the perturbed wave function itself and physical properties would not be changed discontinuously at the core region. Therefore, F_V is a more accurate way to predict forces in the solids.

Figure 4.6 a, b show ρ and C_{11} images in K(100) plane under 0% strain condition, respectively. Lighter contrast expresses a higher absolute value of each physical quantity. Since the probability density of physical quantities is affected from both ρ and S, differences between ρ and other physical quantities, such as ι, ES and ESP, are believed to be caused by the interference of the phase quantity. The difference between ρ and C_{11} in K is caused by this. Although in former studies, the bond between atoms was predicted from ρ, the high ρ region does not always represent the existence of a strong bond. ESP is more accurate way to predict bond image in order to appreciate mechanical properties.

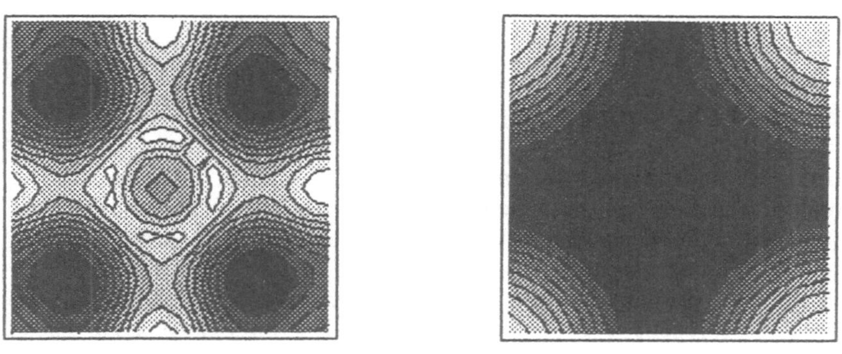

(a) (b)

Fig. 4.5. (100) plane Image of **(a)** F_V and **(b)** F_f under 0% strain in K.

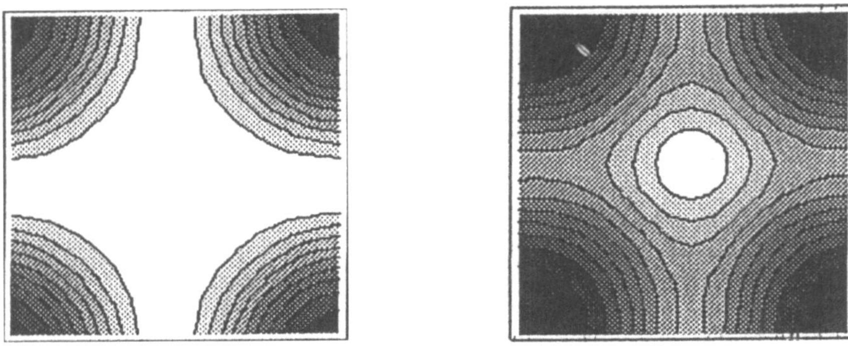

(a) (b)

Fig. 4.6. (100) plane Image of (a) ρ and (b) C_{11} under 0% strain in K.

5. Discussion

5.1 Catastrophic Change of ESP during Tensile Deformation

During tensile deformation over ε_{cr}, ESP shows a catastrophic change from positive to negative values. These critical phenomena are believed to have a certain relationship with ideal fracture. It is curious to note that, not like a partial instability, ESP shows catastrophic transformation in the whole region. Since ESP is the finite difference of ES with strain, negative divergence of ESP means a sudden decrease of ES during tensile elongation.

From this result, a strong dependence of ε_{cr} upon materials is observed, especially for a metal or a semiconductor, which shows good agreement with fracture toughness of the materials. General lattice instability in ideal fracture could be estimated from the concept of stiffness instability; divergence of ESP.

5.2 Change of ESP during Compressive Deformation

Compared with tensile deformation, ESP shows a gradual increase and no catastrophic transition during compressive deformation, which shows good agreement with general mechanical properties. However, in K and Si, certain decreases of ESP to zero are observed in local regions after certain compressive strain. This might relate to plastic deformation, since partial slip planes, which represent a local decrease of bonding force in a solid, cause plasticity. However, discussion of plasticity is beyond the scope of this study.

5.3 Universal Existence of Critical Strain

As shown in Table 4.1, the existence of ε_{cr} is universal and its value is material dependent. From the point of inflection of total energy in the pseudopotential method, Senoo et al.[1] showed the existence of ideal strength under 11% strain in aluminum. Since actual stiffness consists of the ESP term and the Ewald energy term (stiffness from Ewald energy: SEE)[14], and both terms show different critical transformation properties [1], ε_{cr} from ESP and the total energy calculation will be different. However, the critical property of SEE is beyond our discussion. Kitamura [16] also showed critical strain exists in Ni before fracture in MD simulations. Since the point of inflection of energy is primarily deduced from band structure energy (E_{bs}), the radius of pseudo-atom core and the correlation and the exchange energies, ε_{cr} is the structural and material sensitive property. However origins of ε_{cr} are not well understood.

5.4 Origins of Critical Strain

Since ε_{cr} is low in semiconductors and high in metals, certain relations between ε_{cr} and band structure are expected. The principal origins of ε_{cr} might be classified into three factors in band structure deformation; (i) existence of Fermi energy, (ii) accordance between the reciprocal vector of lattice (q) and the node (q_0) of the pseudopotential form factor ($V_{ps}(q)$), (iii) anomalous effect on dielectric function of electron ($\varepsilon(q)$). In the following, each factor will be discussed.

5.4.1 Existing of Fermi Energy in Band Structure. Since the existence of a band gap is one of the principal discontinuous properties in solids, diversity of ε_{cr} might be caused from the band structure of each material. Generally speaking, during tensile deformation, the wave vector of Brillouin zone boundary (k_{bz}) in the elongation direction is shrunk, whereas k_{bz} is unchanged in perpendicular directions. During deformation, since the electron density of the solid is decreased homogeneously, k_f could reach k_{bz}. Senoo and co-workers[17,18] explain the origin of fracture toughness from the existence of the Fermi surface, where electrons possibly change state with deformation. In semiconductors, since there exists no Fermi surface, ESP catastrophically changes with small strain.

The fact that the existence of a Fermi surface is key for the critical change of ES and ESP reminds us of HOMO and LUMO for chemical reactions in frontier orbital theory. However, since ESP is the summation of whole electrons, an explanation of catastrophic transition of ESP by electrons near the band gap might not be sufficient.

5.4.2 Relation between Structure Factor and First Node of Form Factor.

According to the second order energy of perturbation theory, E_{bs} is expressed with the absolute value of the structure factor ($S(q)$) and energy-wave-number characteristic function ($\Phi_{bs}(q)$) in (5.1). Φ_{bs} is expressed with the exchange-correlation enhanced function ($\chi(q)$, $\varepsilon(q)$, mean volume of electron (Ω) and the absolute value of $V_{ps}(q)$ in (5.2). q_0 is the point, where $V_{ps}(q_0) = 0$. Since q_0 of $V_{ps}(q)$ depends on the core radius of the pseudopotential (R_M) in (5.3), q_0 is independent of the deformation.

During tensile deformation, q will be shrunk in the elongation direction. When q coincides with q_0, one term of (5.1) shows a local maximum. Although E_{bs} is the summation of $|S(q)|^2 \Phi_{bs}(q)$, since generally $S(q_0)$ is larger than the other $S(q)$, structural instability might occur. Although the positional relation between q and q_0 is material dependent, it will not always coincide with diversified K_{Ic} in metals and semiconductors.

$$E_{bs} = \Sigma\,|S(q)|^2\,\Phi_{bs}(q) \tag{5.1}$$

$$\Phi_{bs}(q) = 1/2\,\Omega\,[\chi(q)/\varepsilon(q)]\,|V_{ps}(q)|^2 \tag{5.2}$$

$$q_0 = \pi / 2 R_M \tag{5.3}$$

5.4.3 Anomaly Effect on Dielectric Function of Electron.

The anomaly effect of $\varepsilon(q)$ has been discussed as one of the origins for soft mode of lattice vibration to explain the Kohn anomaly and the phase transformation. The derivative of $\varepsilon(q)$ with reciprocal vector is discontinuous at $2k_f$. However, as the relation between q and k_f, the anomaly of $\varepsilon(q)$ does not directly relate to the difference between metals or semiconductors.

To clarify the origin of ε_{cr}, ESP on each level of electron and the relation between k_f, q_0 and q should be analyzed in reciprocal space.

5.5 Relation between Fracture Mechanics and Critical Strain

Griffith's theorem is based on the energy balance between elastic energy (E_e) and surface energy (E_s). Orowan developed Griffith theorem concerning plastic energy (E_p) as shown in (5.4).

$$E_e > E_s + E_p \tag{5.4}$$

Since E_s is not diversely changed, E_p affects ductility and brittleness of materials. According to Orowan's theory, the plastic deformation ability decides fracture toughness of materials, as shown in Fig. 5.1. It describes the stress-strain curves of metals and semiconductors. Recent calculation of fracture mechanics, such as the J-integral, is also based on elastic and plastic energy.

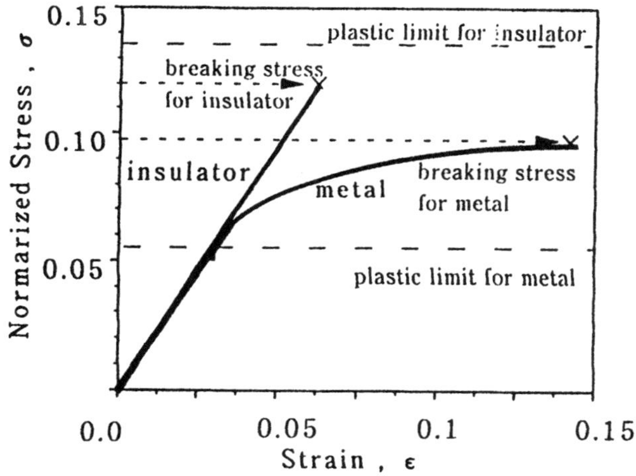

Fig. 5.1. Stress-strain curve and fracture explained by Orowan's theory.

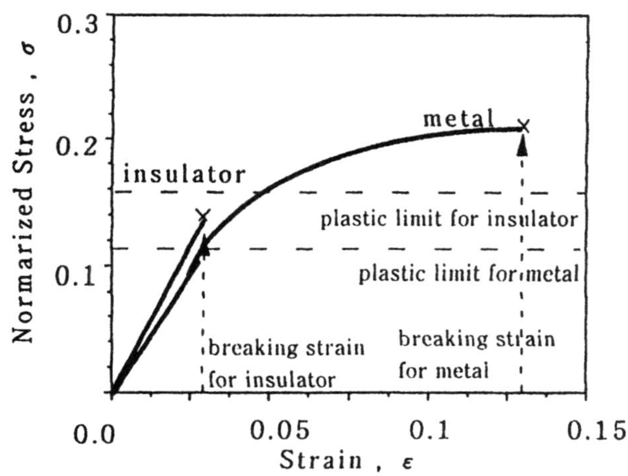

Fig. 5.2. Stress-strain curve and fracture explained by critical strain.

Since ε_{cr} represents the local lattice instability, it might imply the existence of fractural strain in ideal fracture, compared with fractural energy in Griffith's theorem. Although surface energy is environment dependent, ε_{cr} is purely deduced from the state of electrons. ε_{cr} can be one of the parameters to determine the ideal strength of materials. Ideal fracture might be explained in terms of strain, as shown in (5.5), besides energy. Equation (5.5) shows that ideal fracture occurs, when elastic strain (ε_e), which is the difference between total strain (ε_{total}) and plastic strain (ε_p), exceeds ε_{cr}. Figure 5.2 shows stress-strain curves for ductile and brittle materials ruled with ε_{cr} parameters.

$$\varepsilon_{cr} > \varepsilon_e = \varepsilon_{total} - \varepsilon_p \tag{5.5}$$

5.6 Macroscopic, Nanoscopic and Electron Image in Actual Fracture

There is much controversy on the relation between practical fracture and ideal fracture. Under actual conditions, fracture occurs in small strain, whereas ε_{cr} is as much as 30% in metals. In practice, the fractural phenomenon is variously classified into plastic blunting, concentration of dislocations, micro-void coalescence and cleavage. However, in ideally modeled MD calculations, large strain is observed at the local crack initiations [16]. In nanoscopic images of the crack tip region, the strain field is believed to be inhomogeneous and local fracture occurs at the ε_{cr} strained region.

Whereas MD is based on force balance in real space, ES is based on structure change in reciprocal space. It would be interesting to discuss the relationship between the local spatial force and the ESP change in a whole system at the initiation of the lattice instability. Properties of wide spread electrons are believed to cause the catastrophic transformation of ESP in the whole system, unlike breaking of certain bonds.

To connect electron behavior in reciprocal space with practical fracture, which is apparently a local phenomenon in real space, a localized wave image, such as wavelets, and flow of electrons into inhomogeneously strained regions should be considered. Since fracture can be treated as different scales of phenomena, such as macroscopic, nanoscopic and atomistic levels, relations between them will be important.

6. Conclusions

From the calculation of ES and ESP we have studied ideal fractural phenomena and obtained the following results;

(1) Catastrophic transition of ESP was observed in all calculations,

(2) Catastrophic transition of ESP was observed in whole regions, not in a particular bond,

(3) ε_{cr} is material dependent; large in metals but small in semiconductors,

(4) ε_{cr} might have a certain relation with ideal fractural properties.

Acknowledgement

We thank Akio Yasu, undergraduate student in Mie University, for his kind collaboration.

References

1. M. Senoo, I. Fujishiro, M. Hirano: Trans. Jpn. Soc. Mech. Eng. **50**, 744 (1984).
2. P. Ziesche, R. Kaschner: Phy. Rev. B **42**, 9158 (1990).
3. J. Wang, J. Li, S. Yip: Phy. Rev. B **52**, 12627 (1995).
4. W. Pauli: *Handbuch der Physik, Band XXIV, Teil 1*, (Springer, Berlin, 1933) p.88.
5. G. Marc, W. G. McMillan: Adv. Chem. Phys. **58**, 209 (1985).
6. B. M. Deb: Rev. Mod. Phys. **45**, 22 (1973).
7. H. Nakatsuji: J. Chem. Phys. **67**, 1312 (1977).
8. S. W. Ghosh: J. Chem. Phys. **87**, 15 (1987).
9. O. H. Nielsen, R. M. Martin: Phys. Rev. B **32**, 3780 (1985).
10. N. O. Folland: Phys. Rev. B **34**, 8296 (1986).
11. O. H. Nielsen, R. M. Martin: Phys. Rev. B **37**, 10905 (1988).
12. N. O. Folland: Phys. Rev. B **37**, 10908 (1988).
13. J. Grafenstein, P. Ziesche: Phys. Rev. B **41**, 3245.
14. T. Aoki, S. Kotake, Y. Suzuki, M.Senoo: 72th Proc. Jpn. Soc. Mech. Eng. **95**, 178 (1995).
15. S. Kotake, Y. Suzuki, M. Senoo: 72th Proc. Jpn. Soc. Mech. Eng. **95**, 174 (1995).
16. T. Kitamura, *Advances in Material Research*, (Springer-Verlag, Berlin, 1997).
17. M. Senoo, Y. Suzuki, Y. Takahashi, S. Kotake: Materia Japan, **36**, 775 (1995).
18. M. Senoo, Y. Suzuki, S. Kotake: *Proc. IMMM '95*, (Mie Academ.Press, Tsu, 1995) p.605.

An All-electron First-principles Molecular Dynamics Method and a Possibility of its Application to Atomistically Distorted Systems

Kaoru Ohno[1], Yutaka Maruyama[2], Hiroshi Kamiyama[3] Eizo Bei[1], Keiichiro Shiga[1], Zhi-Qiang Li[1], Keivan Esfarjani[1] and Yoshiyuki Kawazoe[1]

[1] Institute for Materials Research, Tohoku University, 2-1-1 Katahira, Aoba-ku, Sendai 980-8577, Japan
[2] National Industrial Research Institute of Nagoya, 1-1 Hirate-cho, Kita-ku, Nagoya 462-8510, Japan
[3] Aomori Public College, 153-4 Goushi-Zawa, Yamazaki, Aomori 030-0134, Japan

Summary. First-principles molecular dynamics (MD) have attracted much current interest as an important and powerful technique to investigate short time behavior of atomic motions in relatively small size systems. Among several first-principles MD techniques, a mixed basis approach, which we have developed in recent years, is applicable to clusters, bulk and surfaces, with all electrons taken fully into account. Here, we present simulations of an atomic collision with C_{60}, and diffusion in an iron crystal.

1. Introduction

First principles techniques have been increasingly used in many areas. In particular, the first-principles molecular dynamics (MD) method has been considered as a powerful technique to study atomic motions in materials. The energetics of defect formation or migration [1], or more dramatically, the kinetics of fracture [2], are challenging topics for applying first-principle MD techniques. Because of the loss of translational symmetry, reliable calculations for defects or fracture are considerably more difficult than standard band structure calculations.

The most commonly used first-principles MD method was invented by Car and Parrinello [3] in 1985. This method, using the pseudopotential plane-wave-expansion technique, has been applied to, e.g. structure optimization [4], hydrogen diffusion in crystals [5] and surface adsorption [6, 7].

In second-row elements, 2p valence orbitals have their net component in the core region and are rather localized, since there is no corresponding 1p core state. Because of these localized orbitals, it is not an easy task to make an efficient pseudopotential for these elements which requires only a small number of plane waves. For the same reason, 3d orbitals in transition metal elements and 4f orbitals in rare-earth metal elements are relatively more localized than other valence orbitals. Therefore, devising efficient pseudopotentials for these elements is not easy.

In 1990, Vanderbilt [8] invented the so-called ultrasoft pseudopotential, which does not require the norm conservation and can reduce the number of plane waves needed in the expansion significantly. This new approach has been applied also to *ab initio* MD simulations for a variety of systems including second row elements and transition metal elements. However, for relatively large systems with defects, surfaces or clusters, a huge number of plane waves are required still.

As for the first-principles MD techniques, there are several alternative approaches studied recently by Japanese researchers, besides the original pseudopotential plane-wave-expansion approach. One is the method using the APW basis which was invented by Oguchi and Sasaki [9]. Another is the method using the mixed basis within the pseudopotential formalism by Kishi and Itoh [10]. The final one is the all-electron mixed basis approach [11], which we have developed in recent years.

As an example of applying this all-electron mixed basis MD approach to the physics of fracture, we mainly investigate a carbon-atom cluster, C_{60}, which has attracted current interest as one of the new nanoscale materials. More specifically, we investigate the collision of a metal atom with a C_{60} molecule, which may be considered as one of the atomistic processes of fracture. Also, in order to demonstrate the feasibility of this method with transition metal elements, we present here recent results on vacancy and interstitial formations, and diffusion in an iron crystal, relevant to the mechanics of fracture.

2. Car–Parrinello Method and its Modifications

Since the paper by Car and Parrinello [3] in 1985, *ab initio* MD techniques have been used widely in a variety of systems. In that paper, the new idea of Car and Parrinello was to integrate with respect to time a second-order (Newtonian-like) equation of motion for the wave functions as well as for atoms. Their very active studies with various applications afterwards have demonstrated the validity and effectiveness of their approach. With this, it is no longer necessary to treat the eigenvalue problem and it is possible to reduce both computational time and memory. The reason for the recent rapid progress in the *ab initio* MD techniques is mainly due to this point.

The basic equation in the Car–Parrinello method is given by the Lagrangian L as a functional of the wavefunction ψ_i (i is a level index of the electronic state) and the atomic position $\mathbf{R}_n$ (n is an atom index). The existence of a Lagrangian guarantees that the system has no energy dissipation, i.e. the total energy is conserved. From the Lagrangian, one may derive the equation of motion for the wavefunctions:

$$\mu\ddot{\psi}_i = -H\psi_i + \sum_j \Lambda_{ij}\psi_j, \tag{2.1}$$

where the Lagrange multiplier Λ_{ij} is used to ensure the orthogonality of the wavefunctions ψ_i. This multiplier is determined so as to orthonormalize the wavefunctions by the Ryckaert method [12].

Using a second order equation for the evolution of the wavefunction leads to an oscillatory behavior of the electronic states around the Born–Oppenheimer (BO) surface. This implies that the computed force between the atoms is not the true physical force. For this reason, the value of the fictitious mass μ in the left-hand-side of (2.1) should be chosen to be rather small so that this electronic oscillation is rapid enough not to affect atomic motions along the BO surface.

In place of (2.1), one may use the first order differential equation called 'the steepest-descent (SD) method', combined with the standard 'Gram–Schmidt orthogonalization technique' which has an advantage of breaking spurious symmetries that may occur in the choice of initial conditions for the electrons. Its basic equation is given by

$$\mu\dot{\psi}_i = -(H - \epsilon)\psi_i, \quad \epsilon = \int \psi_i^\dagger H \psi_i \mathrm{d}\mathbf{r}. \tag{2.2}$$

Note that one has to orthogonalize the wavefunctions at each time step by means of the Gram–Schmidt technique. In (2.2), the Lagrange multipliers are replaced by their expectation value. This replacement is called 'the Payne algorithm' [14]. The SD method is essentially the energetically dissipative algorithm, and is suitable for the determination of the ground state electronic structures of clusters or crystals by the first-principles MD. The advantage of Payne's algorithm is that before calculating the force, the electronic density is converged to its ground state value. The force is therefore 'exact' since it is calculated on the BO surface. Recently, as a much faster convergence technique in place of the SD method, 'the conjugate gradient method' and 'the preconditioned conjugate gradient method' [4] have also been used.

3. All-electron Mixed Basis Approach

It is possible to introduce the mixed basis which incorporates plane waves (PWs) and atomic orbitals (AOs) to describe both the extended and localized characters of wavefunctions. The idea of this mixed basis approach was first pronounced in 1979 by Louie *et al.* [13], who treated crystals composed of 3d elements by this method within a pseudopotential formalism. In their approach, they used Gaussian-type atomic orbitals for describing 3d orbitals.

In order to increase accuracy, we have added core electron orbitals to the basis and made it an all-electron formalism. It is a useful extension to incorporate not only valence atomic orbitals but also core atomic orbitals in the mixed basis approach to be able to handle all electrons instead of just valence electrons.

We have developed two types of the all-electron mixed basis codes; one uses the Slater-type atomic orbitals and the other uses the numerical atomic orbitals. The former code has been applied mainly to the second row elements, while the latter has been developed for transition metal elements as well.

Atomic orbitals were confined inside non-overlapping Wigner–Seitz spheres. This avoids overcompleteness problems and also makes the calculation of the Hamiltonian and overlap matrix elements much easier.

In the *ab initio* MD based on the mixed basis, since the basis is not orthogonal, it is necessary to introduce the overlap matrix S. When one uses the SD technique together with Payne's algorithm [14], the basic evolution equation for the electronic wavefunctions reads

$$\mu S \dot{\psi}_i = -(H - \epsilon_i S)\psi_i, \quad \epsilon_i = \int \psi_i^\dagger H \psi_i \mathrm{d}\mathbf{r}, \tag{3.1}$$

where H is the Hamiltonian of the electronic system. Although this time-evolution equation has the overlap matrix S, one may rewrite it in a form without S by using the Choleski decomposition [13]. That is, introducing the lower triangular matrix U satisfying $S = UU^\dagger$ and defining $\tilde{\psi}_i = U^\dagger \psi_i$ and $H' = U^{-1} H U^{\dagger -1}$, one can derive the standard time-evolution equation without S, which is satisfied by $\tilde{\psi}_i$ and H'. At each time step, however, one has to evaluate the charge density from the expansion coefficients, $\psi_i = U^{\dagger -1}\tilde{\psi}_i$ and the Hamiltonian H from the charge density, and then transform it to H'. Moreover, since the atomic orbitals of the core states are extremely localized, one has to evaluate the charge density, potential, matrix elements and forces analytically or numerically, in the radial direction, as precisely as possible.

The application of the all-electron mixed basis approach to the transition metal clusters [15] or semiconductor crystals [16] has been recently reported. In this method, by systematically incorporating the numerical atomic orbitals into the calculation, it becomes possible to describe precisely the almost spherical core orbitals and to treat the aspherical part of the potential.

4. Alkali-metal Insertion into C_{60}

Endohedral fullerenes have recently attracted considerable interest. Among many investigations, laser vaporization techniques have revealed the production of endohedral C_{60}, in which an alkali-metal atom, like potassium or cesium, is trapped [17]. On the other hand, it has been reported that endohedral $[Li@C_{60}]^+$ and $[Na@C_{60}]^+$ species are formed during collisions of alkali-metal ions with C_{60} vapor molecules [18]. (Here, the notation $Li@C_{60}$, for example, means a LiC_{60} molecule with one Li atom encapsulated inside a C_{60} cage.) Experiments by Ohtsuki *et al.* [20] recently revealed that ^7Be can penetrate into the C_{60} cage to produce $^7Be@C_{60}$ by the recoil of nuclear reactions ($^7Be@C_{60}$ later becomes $^7Li@C_{60}$ by the EC decay). Because of the abundance of C_{60} in the usual fullerene production, endohedral C_{60} should have more practical importance in the future.

In this section, we confine ourselves to the collision between Li^+ and C_{60}^- in a plasma state, where collisions may proceed more effectively due to the electrostatic interaction; and report the result of a constant temperature *ab initio* molecular dynamics simulation yielding an ideal Li@C_{60}. Also, we present briefly our results of the collision between Na^+ and C_{60}^-, and discuss generally the possibility and mechanism of the encapsulation of alkali-metal elements in C_{60} [19].

The electronic states are treated in an all-electron mixed basis approach within the framework of the local density approximation (LDA). The population of the excited levels is determined by the Fermi–Dirac (FD) distribution function. The method of simulation employed here is the same as our previous work concerning the stability and reactivity of C_{60} [11]. In the present simulation, we put one C_{60} molecule and one alkali-metal atom in a supercell with given initial velocities. Once the electronic states are well converged by the steepest descent (SD) method, a charge transfer from the alkali-metal atom to the C_{60} molecule takes place when the Li-C_{60} distance becomes small. In the calculation, we used 60×5 (1s,2s,$2p_x$,$2p_y$,$2p_z$) Slater type atomic orbitals (STOs) and 2969 plane waves corresponding to a cutoff energy of 5.7 Ry. For the damping factors of the 1s and other STOs we used, respectively, $1/0.18a_B$ and $1/0.24a_B$, where $a_B = 0.52918$ Å is the Bohr radius. The supercell was divided into $64 \times 64 \times 64$ meshes and 1 a.u. $= a_B$ corresponded to 2.7 meshes. We set $\Delta t = 4$ a.u. ~ 0.1 fs and performed six SD iterations between two adjacent updations of atomic positions in order to converge the electronic states.

Several different simulations were performed with different initial kinetic energy (KE) of the alkali-metal element, and impact parameter. In all the simulations, the temperature was set to $T = 1000$ K by a velocity rescaling for the atoms and using a FD distribution for the occupation of electronic levels. Below, we show only the conditions assumed in the simulations and the main results.

(A) Li^+ ion with KE $= 5$ eV hits the center of a six-membered ring in C_{60}^- perpendicularly (see Fig. 4.1). In this case, the Li^+ ion penetrates into the cage through the center of the six-membered ring without difficulty, since the ionic radius of Li^+ is 0.6–0.9 Å, and the hole of a six-membered ring is of the same size or slightly smaller. That is, a direct penetration of the Li ion through the center of a six-membered ring occurs with suitably high kinetic energy of the Li^+ ion. The Li^+ ion passes deeply into the C_{60} (at a distance of 1.6 Å from the six-membered ring) and comes back because of the Coulomb interaction resulting from charge transfer, to be trapped at a distance of 1.0 Å from the center of the same six-membered ring. The final position of the Li^+ ion near the center of a six-membered ring is more favorable, because of the

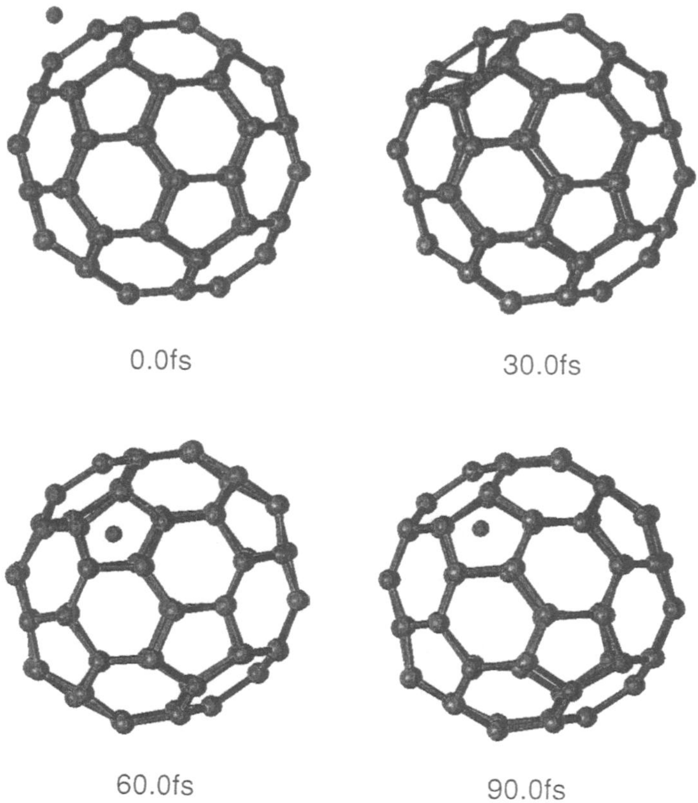

Fig. 4.1. Snapshots of the simulation (A), where Li^+ (*top left*) hits with 5 eV kinetic energy the center of a six-membered ring of C_{60}^-. The figures show only the positions of the carbon atoms in C_{60} and Li atom. In this case, after 60 fs, Li@C_{60} is created.

initial condition of insertion (vertically toward the center of a six-membered ring), and because of the symmetry.

(B) Li^+ ion with KE = 1 eV hits the center of a six-membered ring of C_{60}^- perpendicularly. In this case, the Li^+ ion bounces several times on the cage of C_{60} and ends up at the center of a six-membered ring outside the cage.

(C) Li^+ ion with KE = 5 eV hits the C_{60}^- near a double bond (off center of a six-membered ring) perpendicularly (see Fig. 4.2). In this case, the Li^+ ion bounces several times on the cage of C_{60} and ends up bound to the outside of the cage similarly to (B). By the shock of the initial and successive collisions of the Li^+ ion, the cage deforms near the position of the collision. All these simulations have been carried out for 75–90 fs, thus further annealing is presumably possible.

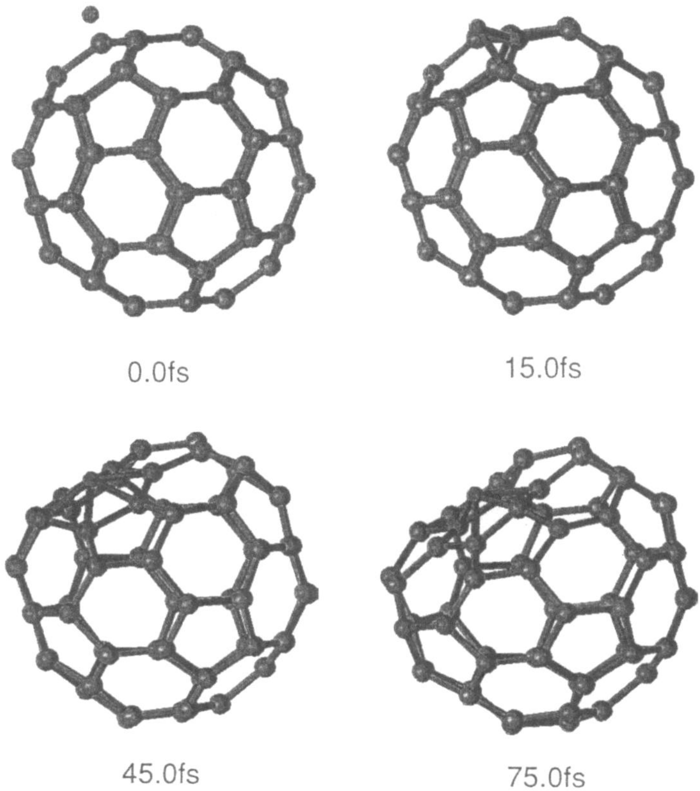

Fig. 4.2. Snapshots of the simulation (C), where Li^+ hits with 5 eV kinetic energy near a double bond of C_{60}^-.

In the real experimental situation, the mean time interval for alkali-metal positive ions with average KE of 5 eV to hit the center of a six-membered ring of C_{60}^- molecules in a direction nearly perpendicular to the cage sphere would be $\sim 10^{-10}$ s, if 10^{11} lithium ions and the same number of C_{60} ions exist in a 100 cm^3 box. Therefore, it is reasonable to consider that the direct insertion process as in simulation (A) is the mechanism of the Li@C_{60} formation. However, the situation might be somewhat different in the case of alkali-metals heavier than Li. The ionic radius of Na^+ is 0.95–1.16 Å, slightly larger than the radius of the hole of a six-membered ring, while the radius of K^+ is 1.33–1.52 Å much too large to enter through the six-membered ring. Therefore the direct insertion process is rather unlikely because of the strong overlap between the electronic clouds of the cage surface and the alkali-metal ion. A possible mechanism of the insertion of a Na^+ ion was discussed in our previous paper as a process with a long time period [19].

The above is an example of mechanical distortion in nanoscale materials, which is considered to be a fundamental atomic procedure occurring in 0.1 ps order of time and first "observed by computer simulation" using the first principles treatment.

5. Vacancy Formation and Diffusion in an Iron Crystal

Now we turn our attention to the problem of diffusion in an iron crystal. As mentioned in the introduction, these calculations are not easy because very high accuracy is required to obtain reliable total energies. We apply the all-electron mixed basis approach with numerical atomic orbitals to such systems. Here we will briefly show some of our recent results.

First, 16 atoms are put inside a cubic unit cell of size 5.73×5.73×5.73 Å^3 to form a perfect bcc lattice. Calculating total energies of this perfect system and a system with a single vacancy (15-atom system), we estimated the vacancy formation energy of iron crystal to be 1.2 ±0.3 eV, which is comparable to experimental values. This calculation was performed with the local spin-density approximation (LSDA). The number of k-points in the irreducible Brillouin zone (1/48 of the first Brillouin zone) is 4 and the PW cut-off energy is 16.5 Ry (the number of PWs is 1419, and the unit cell is divided into 64×64×64 meshes). For atomic orbitals, which are generated by Herman–Skillmann's atomic code, we use the 1s, 2s, 2p, 3s, 3p, 3d and 4s atomic orbitals with the cutoff radius, $r_c = 2.0a_B = 2.0$ a.u..

Next, in order to simulate a self-interstitial diffusion, we put 55 iron atoms in a larger cubic cell of size 8.60×8.60×8.60 Å^3 to form an imperfect bcc lattice. Here also the unit cell is divided into 64×64×64 meshes. The initial speed of atoms is set to be 0. For this system, we use the LDA instead of LSDA, and the 1s, 2s, 2p, 3s, 3p and 3d atomic orbitals together with the 3071 PWs of a cutoff energy of 12 Ry. After the initial 50 iterations to reach the electronic ground state, an MD simulation was performed with a time step of $\Delta t =$ 4 a.u. = 0.1 fs. Typically 6 SD steps were performed to reach the electronic ground state before every atomic update. Starting from four different initial configurations, we optimized the local configurations. Figure 5.1 represents the structure of the <110> dumbbell optimized by the present method. The five bonds drawn in the figure are shorter than the other bonds. These bond lengths are given as follows: The center bond is 1.86 Å, the lower two are 1.87 Å and the upper two are 1.89 Å. Compared to the total energy of the pure single crystal with 54 atoms in the unit cell, we identified the interstitial formation energy to be 2.4 eV, which is slightly smaller than the experimental values. The actual interstitial formation energy experimentally observed can have a somewhat larger value than this, since the interstitial formation requires additionally the activation energy of order 1 eV to reach the optimal interstitial structure. So far, the <110> dumbbell has been considered to have the minimum energy compared to the other interstitial configurations

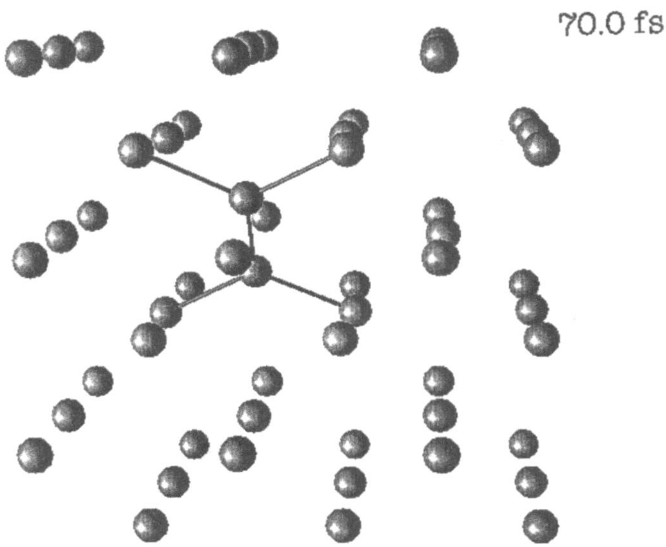

Fig. 5.1. Optimized structure of the <110> dumbbell in bcc iron crystal.

in the bulk [21]. However, our tentative study revealed that this is not always the case. In this kind of simulation, one may assume several possible paths for the diffusion process of the interstitial atom. Assuming the nearest-neighbor diffusion path for the interstitial iron atom from the (110) dumbbell at the cube corner to the (101) dumbbell at the body center, we calculated the total energy of several configurations along this path. In the middle of this path, we accidentally found a local configuration which has the minimum energy, still slightly lower than the <110> dumbbell configuration, although the cell-size dependence, etc. should be more carefully checked before coming to any decisive conclusion. The detail of these results will be reported elsewhere [22].

6. Concluding Remarks

Here, we have presented an introductory description of the first-principles MD techniques and discussed the possibility of applying them to the physics of defects and fracture. The amount of calculations required for determining migration or reaction paths with these methods is, in general, extremely large even when we use the highest performance supercomputers currently available. In this article, we have paid special attention to the all-electron mixed basis approach, which may handle the second-row elements and transition-metal elements with rather small numbers of plane waves compared to the conventional pseudopotential approach. Although this approach is just one possibility of the new formalisms in the first-principle MD, we hope, in the near future, that more realistic larger scale systems can be handled successfully and accurately by using such methods.

7. Acknowledgement

We are grateful to the Materials Information Science Group of the Institute for Materials Research, Tohoku University, for continuous support of the HITAC S-3800/380 supercomputing facilities.

References

1. L. A. Girifalco: '*Atomic Migration in Crystals*' (Blaisdell Pub., New York, 1964).
2. B. L. Averbach, D. K. Felbeck, G. T. Hahn, D. A. Thomas: (eds), '*Fracture*' (MIT Press, Cambridge, Massachusetts, 1959).
3. R. Car, M. Parrinello: Phys. Rev. Lett. **55**, 2471 (1985).
4. As a review, see: M. C. Payne, M. P. Teter, D. C. Allan, T. A. Arias, J. D. Joannopoulos: Rev. Mod. Phys. **64**, 1045 (1992).
5. F. Buda, G. L. Chiarotti, R. Car, M. Parrinello: Phys. Rev. Lett. **63**, 294 (1989).
6. M. Needels, M. C. Payne, J. D. Joannopoulos: Phys. Rev. Lett. **58**, 1765 (1987).
7. S. Ihara, S. L. Ho, T. Uda, M. Hirano: Phys. Rev. Lett. **65**, 1909 (1990).
8. D. Vanderbilt: Phys. Rev. B **41**, 7892 (1990).
9. T. Oguchi, T. Sasaki: in: '*Computer Aided Innovation of New Materials II*', eds. M. Doyama, J. Kihara, M. Tanaka, R. Yamamoto, (North-Holland, Amsterdam, 1993) p.107.
10. T. Kishi, S. Itoh: J. Phys., Condens. Matter **7**, 6197 (1995).
11. K. Ohno, Y. Maruyama, Y. Kawazoe: Phys. Rev. B **53**, 4078 (1996).
12. J. P. Ryckaert, G. Ciccotti, H. J. C. Berendsen: J. Comput. Phys. **23**, 327 (1977).
13. S. G. Louie, K.-M. Ho, M. L. Cohen: Phys. Rev. B **19**, 1774 (1979).
14. M. C. Payne, M. Needels, J. D. Joannopoulos: Phys. Rev. B **37**, 8137 (1988).
15. Y. Maruyama: Ph.D. Thesis, Tohoku University (1996).
16. K. Ohno, F. Mauri, S. G. Louie: Phys. Rev. B **56**, 1009 (1997).
17. Y. Chai *et al.*: J. Phys. Chem. **95**, 7564 (1991).
18. Z. Wan *et al.*: Phys. Rev. Lett. **69**, 1352 (1992).
19. K. Ohno, Y. Maruyama, K. Esfarjani, Y. Kawazoe, N. Sato, R. Hatakeyama, T. Hirata, M. Niwano: Phys. Rev. Lett. **76**, 3590 (1996).
20. T. Ohtsuki, K. Masumoto, K. Ohno, Y. Maruyama, Y. Kawazoe, K. Sueki, K. Kikuchi: Phys. Rev. Lett. **77**, 3522 (1996).
21. H. Kamiyama, R. Raffi-Tabar, Y. Kawazoe, H. Matsui: J. Nucl. Mater. **212/215**, 231 (1994).
22. H. Kamiyama, K. Ohno, Y. Maruyama, Z.-Q. Li, Y. Kawazoe, H. Matsui: (in preparation).

Part 6

New Computational Methods

Pressure Calculation Scheme in a Small Control Volume

Tamio Ikeshoji

National Institute for Advanced Interdisciplinary Research, 1-1-4 Higashi, Tsukuba, Ibaraki 305, Japan . (Present address, Tohoku National Industrial Research Institute, 4-2-1, Nigatake, Miyagino-ku, Sendai 985, Japan)

Summary. A calculation scheme of the pressure tensor in a molecular dynamics simulation is discussed. A general expression of the pressure in a control volume is presented. The virial of a two-atom pair contributes the pressure between the two atoms according to how long their connecting line passes through the control volume. It is shown that this scheme gives a reasonable pressure profile. The pressure perpendicular to an interface is computed to be constant by this scheme throughout the equilibrium system, unlike the artificial hump in the pressure profile, which owns with the widely used method.

Key Words. Molecular dynamics, pressure tensor, control volume

1. Introduction

In molecular dynamics (MD), the pressure tensor, **P**, in a unit cell under the periodic boundary conditions is generally calculated with a virial of the system as

$$\mathbf{P} = -\left\langle \sum_{i}^{unit\ cell} \sum_{j}^{all} \left\{ \boldsymbol{r}_{ij} \boldsymbol{f}_{ij} \right\} \right\rangle / 2V + \delta_{\alpha\beta} k_B T \rho \quad , \tag{1.1}$$

where $\boldsymbol{r}_{ij}$ is the position of particle j relative to i, $\boldsymbol{f}_{ij}$ is the force onto i due to j, V is the volume of the unit cell, T is the temperature, ρ is density, $\delta_{\alpha\beta}$ is Kronecker's delta (α and β indicate directions.), and k_B is Boltzmann constant. Summation is for all i-j pairs in the system. Although this equation is used widely, a problem owns when pressure is computed in a subjection system. The same problem occurs in the pressure tensor calculation in liquid droplets which has generated many discussions on the pressure profile calculation. In the literature, many interesting statements are found; "the stress tensor is not uniquely defined" [1], "there are infinitely many ways of writing **P**($\boldsymbol{r}$)" [2] which can be distinguished by different choices of a contour joining two interacting molecules. Two kinds of contours have been presented [3]; "Irving-Kirkwood definition [4] by the straight line" [5] and "Harasima's expression [4] by choosing an unsymmtric contour" [2].

Since the first paper on the calculation of the pressure at the molecular level appeared in 1950 [4], the issue is still not settled [6-12]. In many studies in this field, statistic mechanics was used to derive the schemes, even though MD calculations can give the forces in a straightforward. In this article, we use a purely mechanical approach. A simple and widely used way to calculate the pressure in a small volume in the system is to sum up the virials due to interacting atoms, of which atoms, at least one of the pair, are located in the control volume (a local volume in the system), CV, as follows.

$$\mathbf{P} = -\left\langle \sum_{i}^{CV} \sum_{j}^{all} \left\{ \boldsymbol{r}_{ij} \boldsymbol{f}_{ij} \right\} \right\rangle / 2V_{cv} + \delta_{\alpha\beta} k_B T \rho_{cv} \tag{1.2}$$

The last term may be correct, since it is the ideal pressure term which comes from how often atoms hit or pass through walls in the control volume. The first term comes from the force through the walls. This term in the above equation contains some problems. Assume a small volume and that two atoms which interact with each other but are outside of this volume. According to (1.2), the virial due to this pair does not contribute the pressure in this small volume. It looks to be correct to exclude the virial of the atom pair outside of the volume, since only atoms feel the forces. However, the virial does not represent a pressure on a certain surface.

2. Pressure of the Whole System

In this section, the method of calculating the pressure of the whole system is described briefly. It can be found in many text books e.g. [13].

2.1 Pressure by Mechanical Way

Suppose an infinitely large plane in the system (See Fig. 1a). Left side atoms in the unit cell feel forces from atoms in the right side by

$$F^{\text{left}} = -\left\langle \sum_{i \in right}^{unit\ cell} \sum_{j \in left}^{all} \left(\boldsymbol{e}_n \boldsymbol{f}_{ij} \right) \right\rangle, \tag{2.1}$$

where $\boldsymbol{e}_n$ is the unit vector perpendicular to the plane. The same force with the opposite sign will be observed on the right side. Therefore, the pressure due to these forces on this plane is

$$P^{force} = F^{\text{left}} / S\text{,} \tag{2.2}$$

where S is a surface area of the plane within the unit cell. When we scan this plane from the edge of the unit cell to another edge and average the pressure observed on the plane, the following expression for the pressure of the whole system is obtained.

$$P^{force} = -\left\langle \sum_{i}^{unit\ cell} \sum_{j}^{all} \left(\boldsymbol{e}_n \boldsymbol{r}_{ij}\right)\left(\boldsymbol{e}_n \boldsymbol{f}_{ij}\right) \right\rangle / 2V \tag{2.3}$$

This is the origin of the virial that appeared in the expression for the pressure. In this derivation, an imaginary plane is inserted between two interacting atoms. We defined the pressure on it with the atoms which feel the forces. The plane does not feel the forces.

2.2 Pressure by Volume Work

When the system is uniformly compressed by α in the n-direction, the distance between an i-j pair will decrease by $\alpha|\boldsymbol{e}_n \boldsymbol{r}_{ij}|$. Therefore, the potential energy of the system will increase by

$$\sum_{i}^{unit\ cell} \sum_{j}^{all} \left(\alpha \boldsymbol{e}_n \boldsymbol{r}_{ij}\right)\left(-\boldsymbol{e}_n \boldsymbol{f}_{ij}\right) / 2 \tag{2.4}$$

This energy increase corresponds to the work $PS(\alpha V/S)$ done by the pressure to move surface of S in the n-direction by $\alpha V/S$. Then, the same equation as (2.3) is obtained. This is the derivation of the virial by the work.

2.3 Ideal Gas Pressure

The second term in (1.1) is obtained by considering how atoms pass through the plane in the unit cell as explained for the pressure by the ideal gas kinetics.

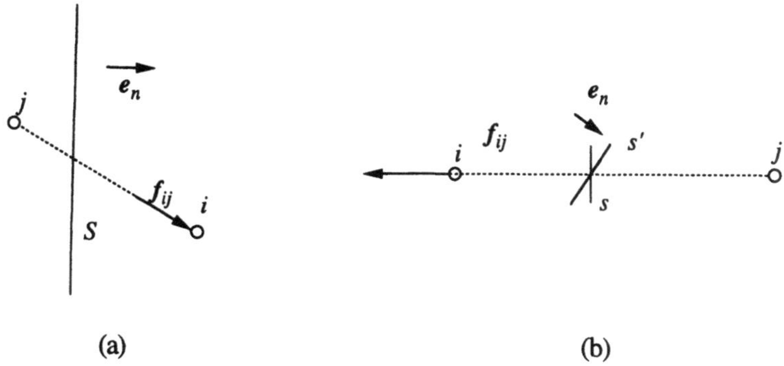

Fig. 1 . Pressure due to i-j pair, (**a**) on infinite surface S in the infinite or in the unit cell under the periodic boundary conditions (**b**) on small surface s in the control volume.

3. Pressure in Control Volume

Many ways for calculation of the pressure tensor in a small volume in the system have been presented [6-12]. A new algorithm, which is suitable to use for plane layers in MD calculation, is described here. It can be easily extended for spherical layers.

3.1 A General Expression of Pressure in Control Volume

When we considered the whole system, it was possible to obtain the pressure on the infinitely large plane. However, it is impossible to put such a large plane in a small volume to consider the pressure on it. It is also difficult to calculate the pressure work to change the control volume, since it is impossible to change only the control volume. The remainder will also change. So, let us extend the first idea to calculate the pressure in the volume. What will happen when we put a smaller size of plane than the control volume? The same contribution to the pressure may be observed, if the force acts through the plane. It can be arbitrarily defined how the force acts as described in the introduction [1, 3]. The most straightforward definition is the straight line between i and j by Irving and Kirkwood [4]. Another force acting line by Harasima may be suitable only for curved and plane layers [1, 6]. In this note, we use the same straight line as Irving-Kirkwood.

When a plane of surface area s is put between atoms i and j as shown in Fig. 1b, the pressure observed on it is $-f_{ij}/s$, if all the force passes through this plane. The pressure on the plane, s', which is towards $\boldsymbol{e}_n$, is, then:

$$-\boldsymbol{e}_n\boldsymbol{f}_{ij} / s' = -\left(\boldsymbol{e}_n\boldsymbol{r}_{ij}\right)\left(\boldsymbol{e}_n\boldsymbol{f}_{ij}\right) / r_{ij}\, s \qquad (3.1)$$

Next, we scan the plane from position a to b on the i-j connecting line. The positions of a and b are at the edges of the control volume, or position of the atoms if they are in the target control volume. The average pressure observed during scanning may be calculated by integration with weight of the surface area s and considering the contribution to the control volume from the scanned volume as

$$P_n = -\left(1 / r_{ij}\, V\right) \int_a^b \left(\boldsymbol{e}_n\boldsymbol{r}_{ij}\right)\left(\boldsymbol{e}_n\boldsymbol{f}_{ij}\right) dq + k_B T \rho_{cv}, \qquad (3.2)$$

where q is the coordinate on the i-j connecting line (from i to j). This is the basic equation of the pressure in the control volume [15]. This equation is in the same form as the following basic equation proposed by Irving and Kirkwood [4].

$$P_{\alpha\beta}(\boldsymbol{r}) = \frac{1}{2}\left\langle \frac{r_{ij,\alpha}}{r_{ij}} f_{ij} \int_{C_{ij}} dl\, \delta(\boldsymbol{r} - \boldsymbol{l}) \right\rangle + \delta_{\alpha\beta} k_B T \rho_{cv}, \qquad (3.3)$$

where C_{ij} is any contour connecting i to j, and $\delta(\boldsymbol{r}$-$\boldsymbol{l})$ is the Dirac delta function.

3.2 Calculation Scheme of the Pressure in Flat Layers

In the case that $\boldsymbol{e}_n$ in (3.2) is constant through the integration line like the pressure tensor of x-, y-, and z-directions in a sublayer, the pressure in n-direction becomes simply the virial expression as

$$P = -\frac{\Delta q}{r_{ij}} \sum_i^{CV} \sum_j^{all} \left(\boldsymbol{e}_n\boldsymbol{r}_{ij}\right)\left(\boldsymbol{e}_n\boldsymbol{f}_{ij}\right) / 2V_{cv}, \qquad (3.4)$$

where Δq is the distance which the i-j connecting line passes through. Even in the case that either the i nor the j atom is in the layer, there is a contribution to the pressure in that layer if the i-j connecting line passes through the layer. This does not mean that the atoms in the layer feels the force from the i-j pair of which connecting line it completely passes through.

3.3 Calculation Results of the Pressure through the Lattice Boundary

The pressure profile through the lattice boundary was calculated by the two equation. The result by (1.2) showed a strange hump in the x-direction as shown in Fig. 2a, even though the equilibrium was satisfied. In this calculation, two lattices of total 672 Lennard-Jones particles (12-6 potential with potential depth ε and atomic diameter σ) were placed in the unit cell (10.7σ x 6.5σ x 7.6σ) in the three-dimensional periodic boundary conditions. The lattice boundary was perpendicular to the x-coordinate. The [111] directions of the center and side lattices were in y- and x-directions, respectively. If this hump were real, the system will move with no force balance. When (3.4) is used, such a strange hump completely disappears, showing the constant pressure in the x-direction (Fig. 2b). Nonuniform pressures observed in both cases in the y- and z-directions are due to the surface tension at the heterogeneous interface.

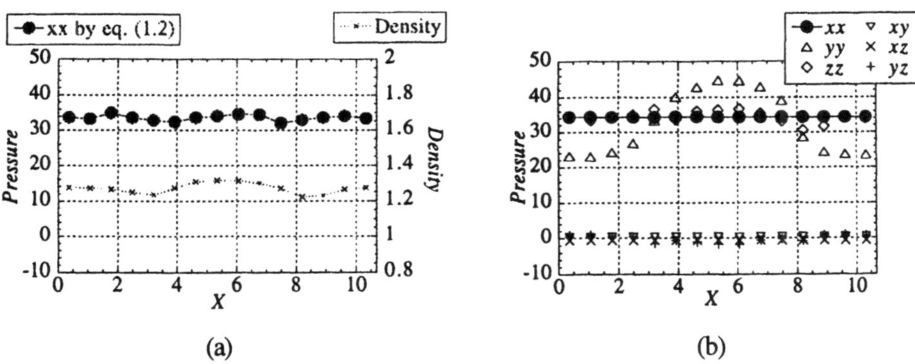

Fig. 2. Pressure profile through the two fcc boundary at $T = 0.6$.
(**a**) from (1.2). (**b**) from (2.8).

4. Concluding Remarks

Pressure in a small part in the system could be calculated to attribute the virial of the two interacting atoms to the small volume in which the connecting line passes. In the case of the pair potential, the equation (3.2) is well defined.

However, there is a problem in the case of many body forces and *ab initio* forces, where virial of $\boldsymbol{r}_i \boldsymbol{f}_i$ can be used to calculate the pressure in the whole system. But, it is difficult to attribute the virial to a small control volume.

References

1. P. Schofield, J.R. Henderson; Proc. R. Soc. Lond. **A379**, 231-246 (1982).
2. J.S. Rowlinson, B. Widom; *Molecular Theory of Capillarity* (Clarendon press, Oxford, 1982).
3. S.M. Thompson, K.E. Gubbins, J.P.R.B. Walton, R.A.R. Chantry, J.S. Rowlinson; J. Chem. Phys. **81**, 530-542 (1984).
4. J.H. Irving, J.G. Kirkwood; J. Chem. Phys. **18**, 817 (1950).
5. E.M. Blokhuis, D. Bedeaux; J. Chem. Phys. **97**, 3576 (1992).
6. A. Harasima; Adv. Chem. Phys. **1**, 203 (1952).
7. M.L.P. Nijmeijer, A.F. Bakker, C. Bruin, J.H. Sikkenk; J. Chem. Phys. **89**, 3789-3792 (1988).
8. M.J.P. Nijmeijer, C. Bruin, A.B. van Woerkom, A.F. J. Bakker, M.J. van Leeuwen; J. Chem. Phys. **96**, 565-76 (1992).
9. E.N. Brodskaya, A.I. Rusanov, M.J.P. Nijmeijer, C. Bruin, J.M.J. Van Leeuwen; J. Chem. Phys. **100**, 7844-5 (1994).
10. E.M. Blokhuis; D. Bedeaux; *HCR Advanced Education* Rev. 56-67 (1994)
11. J.S. Rowlinson; J. Phys. Condensed Matter, A1-A8 (1994).
12. R. Lovett, M. Baus; J. Chem. Phys. **106**, 635-644 (1997).
13. M. Mareschal, M. Baus, R. Lovett, J. Chem. Phys. **106**, 645-654 (1997).
14. J.M. Haile; *Molecular Dynamics Simulation* (John-Wiley, New York, 1992), pp. 333-339.
15. T. Ikeshoji, B. Hafskjold, H. Furuholt (to be submitted).

Accelerated Molecular Dynamics Method

Masuhiro Mikami

National Institute of Materials and Chemical Research, Tsukuba, Ibaraki 305, Japan,
e-mail : mikami@ccmail.nimc.go.jp

Summary. We present an accelerated molecular dynamics method combined with the fast multipole method and the multiple time step integration method for large scale systems with up to a million atoms which interacts with the van der Waals, hydrobonding and electrostatic interaction in isobaric-isothermal (NTP) ensembles.

Key Words. Molecular dynamics method, fast multipole expansion method, multiple time step integration method, potential function, materials science

1. Introduction

During the past forty years, the molecular dynamics method has been widely applied to various materials with thousands of atoms such as molecular crystals, amorphous polymers, liquid crystals, zeolite, nucleic acid, proteins, lipid bilayer, superionic conductors and semi-conductors, because of the advancement of computational algorithms and computer technology. At present, we can predict the crystal structures, thermodynamic properties, and elastic constants of materials which consist of the few thousand atoms, if we only had accurate interatomic potentials for them. However, we have the following difficulties with the molecular dynamics method for mesoscopic systems, which consists of a million of atoms. It is difficult to perform a long time molecular dynamics simulation for a large scale system because of the large computational time associated with the calculation of long range interactions.

Recently, Greengard and Rokhlin [1] developed the fast multipole method (FMM) which is one of a class of algorithms called the hierarchical tree method. The major difference between the FMM method of Greengard and Rokhlin and other tree methods is in the way the far-field interactions are handled. Then, the FMM method makes the computational efficiency essentially O(N), where N is the number of atoms. In addition to its computational efficiency, the error in the FMM can be reduced to close to the machine accuracy by keeping enough terms in the multipole expansion. Schmidt and Lee [2] have also developed the FMM method in a three-dimensional system with periodic boundary conditions. Ding et al. [3] have applied the FMM method to realistic polymer systems with up to 1.2 million atoms.

Furthermore, explicit reversible multiple time step integrators have been developed to handle efficiently problems involving stiff vibrations, disparate masses and long-range forces that occur in large-scale molecular dynamics calculations such as those by Tuckerman et al. [4]. The methods are based on the Liouville operator formation of Hamiltonian mechanics. Procacci and Berne [5] have developed the multiple time molecular dynamics method for the NHP ensemble, and Martyna et al. [6] have also extended this integration method to the NTP ensemble.

In this article, we describe a new accelerated molecular dynamics method combined with the multiple time step integration method and fast multipole expansion method for the NTP ensemble system. In Sect. 2, we outline the equation of motion for the NTP ensemble system, in Sect. 3, we also outline the FMM by Greengard and Rokhlin, in Sect.4, we describe the multiple time step integration method, and show how to combine the FMM and the multiple time step integration method, in Sect. 5, we discuss applications and in Sect. 6 summarize this article.

2. Equation of Motion

The equation of motion for the isothermal-isobaric (NTP) ensemble is established by Andersen [7], Parrinello and Rahman [8], Nose [9], Hoover [10]. After that, Martyna et al. [11] employed a chain of Nose-Hoover thermostats to drive a dynamical system to generate canonically distributed positions and momenta. Also, Martyna et al. [12] proposed to employ the symmetrized pressure tensor in the equation of motion to eliminate the MD cell rotation. The Hamiltonian of the dynamical system including such modifications is given by (2.1).

$$H=\sum_{i=1}^{N}\frac{\vec{p}_i^{\,2}}{2m_i}+\frac{1}{2W}\mathrm{Tr}\left[\vec{p}_h^{\,t}\vec{p}_h\right]+\sum_{j=1}^{M}\frac{p_{\xi_j}^2}{2Q_j}+\Phi(\vec{r},\vec{h})+P_{ext}\det\left[\vec{h}\right]+ \\ (3N+d^2)k_BT\xi_1+k_BT\sum_{j=2}^{M}\xi_j \tag{2.1}$$

where $\vec{p}_i,\vec{r}_i,m_i$ are, respectively, the momentum vector, position vector and mass of atoms, $\vec{p}_h^{\,t},\vec{h},W$ are the cell variable momentum matrix, the cell matrix $(V=\det[\vec{h}])$ and the virtual mass of the MD (molecular dynamics) cell, which are, respectively, the momentum, variable and virtual mass of heat bath, N is the total number of atoms in the MD cell, M is the number of the heat baths, P_{ext},T are respectively, external pressure and temperature, d is the dimension of the MD system, $\Phi(\vec{r},\vec{h})$ is the potential energy function. The equations of motion for the isothermal-isobaric (NTP) ensemble are derived from the Hamiltionian (2.1).

$$\dot{\vec{r}}_i = \frac{\vec{p}_i}{m_i} + \frac{\vec{p}_h}{W}\vec{r}_i$$

$$\dot{\vec{p}}_i = \vec{F}_i - \frac{\vec{p}_h}{W}\vec{p}_i - \frac{1}{3N}\frac{\mathrm{Tr}[\vec{p}_h]}{W}\vec{p}_i - \frac{p_{\xi_1}}{Q_1}\vec{p}_i$$

$$\dot{\vec{h}} = \frac{\vec{p}_h\vec{h}}{W}$$

$$\dot{\vec{P}}_h = V(\vec{P}_{\mathrm{int}} - P_{ext}\vec{I}) + \left[\frac{1}{3N}\sum_{i=1}^{N}\frac{\vec{p}_i^{\,2}}{m_i}\right]\vec{I} - \frac{p_{\xi_1}}{Q_1}\vec{p}_h$$

$$\dot{\xi}_j = \frac{p_{\xi_j}}{Q_j}$$

$$\dot{p}_{\xi_1} = \sum_{i=1}^{N}\frac{\vec{p}_i^{\,2}}{m_i} + \frac{1}{W}\mathrm{Tr}[\vec{p}_h^{\,t}\vec{p}_h] - (3N+d^2)k_BT - p_{\xi_1}\frac{p_{\xi_2}}{Q_2}$$

$$\dot{p}_{\xi_j} = \frac{p_{\xi_{j-1}}^2}{Q_{j-1}} - k_BT - p_{\xi_j}\frac{p_{\xi_{j+1}}}{Q_{j+1}}$$

$$\dot{p}_{\xi_M} = \frac{p_{\xi_{M-1}}^2}{Q_{M-1}} - k_BT. \qquad (2.2)$$

where $\vec{I}$ is a unit matrix, $\vec{P}_{\mathrm{int}}$ is the pressure tensor given by (2.3).

$$(\vec{P}_{\mathrm{int}})_{\alpha\beta} = \frac{1}{2V}\left\{\left[\sum_{i=1}^{N}\frac{(\vec{p}_i)_\alpha(\vec{p}_i)_\beta}{m_i} + (\vec{F}_i)_\alpha(\vec{r}_i)_\beta - (\vec{\Phi}'\vec{h}^{\,t})_{\alpha\beta}\right] + \left[\sum_{i=1}^{N}\frac{(\vec{p}_i)_\beta(\vec{p}_i)_\alpha}{m_i} + (\vec{F}_i)_\beta(\vec{r}_i)_\alpha - (\vec{\Phi}'\vec{h}^{\,t})_{\beta\alpha}\right]\right\}$$

$$\vec{F}_i = -\frac{\partial\,\Phi(\vec{r},\vec{h})}{\partial\,\vec{r}_i}$$

$$\vec{\Phi}' = \frac{\partial\,\Phi(\vec{r},\vec{h})}{\partial\,(\vec{h})_{\alpha\beta}}. \qquad (2.3)$$

3. Fast multipole expansion method

Current methods remain inadequate to simulate the millions of atoms such as micelle formation, viscoelasticity of polymers, phase transition of liquid crystals etc. In order to carry out simulations for such systems, it is essential (i) to eliminate computational steps that depend quadratically (N^2) on size, (ii) to construct algorithms that scale linearly in N. The fast multipole expansion method is one of order N methods for the calculation of electrostatic potential energy. This method divides the MD cell into a hierarchy of rectangular cells (many levels of cells) shown in Figure 1.

Consider an atom i in cell C_0 (the hatched cell in Figure 1). The interaction with any atoms in those far cells (the gray zone in Figure 1) are computed via the multipole expansion, which will be referred to as far field. The interaction with atoms in the 27 nearby cells cannot be computed through the multipole expansion because they are too close, thus they are computed directly pair-by-pair. These interactions will be referred to as near fields. Hence, potential is decomposed as:

$$V(\vec{r}_i) = \sum_{j\in \text{near}} \frac{q_j}{|\vec{r}_i - \vec{r}_j|} + \sum_{A\in \text{far}} V_A^{pol}(\vec{r}_i - \vec{R}_A) \tag{3.1}$$

where $\vec{r}_i$ is the position vector of atom i outside the cell A, $\vec{R}_A$ is the center of cell A, the sum over A is through all far cell, $V_A^{pol}(\vec{r}_i - \vec{R}_A)$ is the multipole potential expanded about the center of the cell A.

$$V_A^{pol}(\vec{r}_i - \vec{R}_A) = 4\pi \sum_{l=0}^{\infty} \sum_{m=-l}^{l} \frac{M_{lm}^A}{2l+1} \frac{Y_{lm}(\theta_{iA}, \phi_{iA})}{|\vec{r}_i - \vec{R}_A|^{l+1}} \tag{3.2}$$

where M_{lm}^A is the multipole expanded about the center of the cell A, θ_{iA}, ϕ_{iA} are, respectively, the polar angle and the angle of longitude.

$$M_{lm}^A = \sum_{j\in A} z_j r_{jA}^l Y_{lm}^*(\theta_{jA}, \phi_{jA}) \tag{3.3}$$

where $\vec{r}_{jA} = \vec{r}_j - \vec{R}_A$, $\vec{r}_j$ is the position vector of atom j in the cell A.

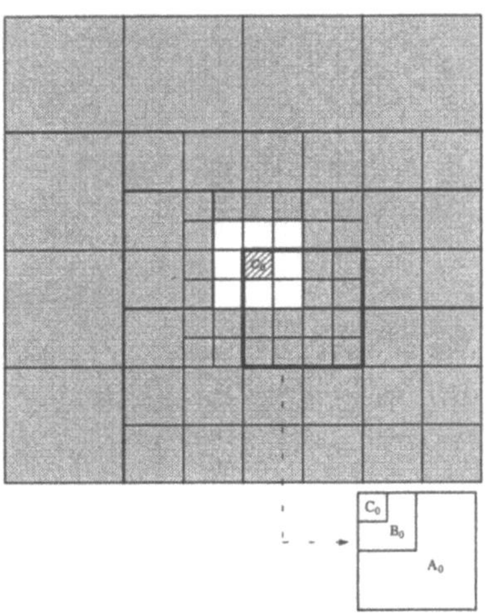

Fig. 1. Schematic diagram for a interaction in 3-levels cell hierarchy. The gray zones are the far cells, the white cell and hatched cells are the near cells.

The algorithm of the FMM method is as follows:

(1) ***Dividing the MD cell into a hierarchy of uniform cells.***

The MD cell is considered as a cell at level 0. This cell is divided into 8 children cells (level 1) each of which is further divided into 8 grandchildren cell (level 2), etc. Finally, at the deepest level L , there are $M = 8^L$ cells.

(2) ***Compute multipole moments for the deepest cells.***

Multipole moments for the deepest level are calculated using (3.3).

(3) ***Compute multipole moments for each cell at each level(the upward pass).***

The parent's multipole moments at level n-1: $M^{(n-1)}_{l'm'}$ are calculated from the children's multipole moment at level n : $M^{(n)}_{lm,k}$ by means of the formula (3.4).

$$M^{(n-1)}_{l'm'} = \sum_{k=1}^{8}\sum_{l=0}^{\infty}\sum_{m=-l}^{l} T^{MM}_{l'm'lm,k} M^{(n)}_{lm,k} \tag{3.4}$$

where $T^{MM}_{l'm'lm,k}$ is the transformation matrix which moves the expansion center of the k-th children's multipoles to the center of the parent's cell.

$$T^{MM}_{l'm'lm,k} = 4\pi \frac{(2l'+1)(-r_k)^{l'-1} Y^*_{l'-l,m'-m}(\theta_k,\phi_k) a_{l'-l,m'-m} a_{lm}}{(2l+1)[2(l'-l)+1] a_{l'm'}} \tag{3.5}$$

$$a_{lm} = (-1)^{l+m} \frac{(2l+1)^{1/2}}{[4\pi (l+m)!(l-m)!]^{1/2}} \, . \tag{3.6}$$

(4) ***Compute the local fields level by level(the downward pass).***

The local field for A_0 which is the grandparent of the cell C_0 is computed from interaction from far cells A in Fig. 1 as

$$V_{A_0}(\vec{r}_i) = \sum_{A} V^{pol}_{A}(\vec{r}_i - \vec{R}_A) \tag{3.7}$$

where $\vec{r}_i$ is the position vector of atom i inside the cell C_0, $\vec{R}_A$ is the center of cell A in the origin of the center of the cell A_0, and the sum over A is through all A cells which excludes the near A cells of A_0 and $V^{pol}_A(\vec{r}_i - \vec{R}_A)$ is given by (3.8).

$$V^{pol}_{A}(\vec{r}_i - \vec{R}_A) = 4\pi \sum_{l'=0}^{\infty}\sum_{m'=-l'}^{l'} L^{(n-1)}_{l'm'} r^{l'}_{iA} Y_{lm}(\theta_{iA},\phi_{iA}) \, . \tag{3.8}$$

The local moments $L^{(n-1)}_{l'm'}$ are computed from the multipole moments of the cell A $M^{(n-1)}_{lm}$.

$$L_{l'm'} = \sum_{l=0}^{\infty}\sum_{m=-l}^{l} T^{LM}_{l'm'lm} M_{lm} \tag{3.9}$$

$$T^{LM}_{l'm'lm} = 4\pi \frac{(-1)^{l+m} Y^*_{l'+l,m'+m}(\theta_t,\phi_t) a_{l'm'} a_{lm}}{r_t^{l'+l+1}(2l+1)(2l'+1) a_{l'+l,m'+m}} \tag{3.10}$$

where $T^{LM}_{l'm'lm,k}$ is the transformation matrix which converts the multipole moments to the local moments at the same level.

Next, the local field for B_0 which is the parent of cell C_0 is computed from interaction from far B cells as

$$V_{B_0}(\vec{r}_i) = V_{A_0}(\vec{r}_i + \vec{R}_{A_0B_0}) + \sum_B V_B^{pol}(\vec{r}_i - \vec{R}_B). \tag{3.11}$$

The sum over B is through all B cells which excludes the near B cells of B_0 and the local field for A_0: $V_{A_0}(\vec{r}_i)$ is shifted by $\vec{R}_{A_0B_0}$ to the center of B_0 by means of the transformation matrix which shifts the expansion center of local moments from the center of A_0 to the center of B_0.

$$L'_{l'm'} = \sum_{l=0}^{\infty} \sum_{m=-l}^{l} T_{l'm'lm}^{LL} L_{lm} \tag{3.12}$$

$$T_{l'm'lm}^{LL} = 4\pi \frac{r_t^{l-l'} Y_{l-l',m-m'}^{*}(\theta_t, \phi_t) a_{l-l',m-m'} a_{l'm'}}{(2l'+1)[2(l-l')+1] a_{lm}}. \tag{3.13}$$

Also, the local field for C_0 is computed from interaction from far C cells in the same manner as the local field for B_0.

$$V_{C_0}(\vec{r}_i) = V_{B_0}(\vec{r}_i + \vec{R}_{B_0C_0}) + \sum_C V_C^{pol}(\vec{r}_i - \vec{R}_C) \tag{3.14}$$

In the deepest level, the interaction potential energy for atom i is computed as

$$V_i(\vec{r}_i) = V_{C_0}(\vec{r}_i) + \sum_{j \neq i} \frac{q_j}{|\vec{r}_i - \vec{r}_j|} \tag{3.15}$$

where the sum over j is through all atoms located in the cell C_0 and its near cells.

4. Multiple Time Step Method

The time evolution equation of the dynamic system which consists of N atoms is given by (4.1).

$$\begin{aligned} &\Gamma(t) = \exp(iLt)\Gamma(0) \\ &iL = \dot{\Gamma} \cdot \nabla_{\Gamma} \\ &\Gamma = \{\vec{r}_i, \vec{v}_i\}, \quad i = 1 \sim N \end{aligned} \tag{4.1}$$

where L is the Liouville operator and Γ is the multidimensional vector of independent variables (coordinates and velocities). In general, the action of the evolution operator on the coordinates cannot be performed analytically. Therefore, a short-time approximation to the true operator, accurate at time $\Delta t = t / P$, is applied P times in succession to evolve the system :

$$\Gamma(t) = \prod_{n=1}^{P} \exp(iL\Delta t)\Gamma(0). \tag{4.2}$$

Also, the time evolution operator is decomposed as (4.4) by means of Trotter's theorem, if the Liouville operator is divided into M_t terms as (4.3).

$$L = \sum_{s=1}^{M_t} L_s \tag{4.3}$$

$$e^{iL\Delta t} = e^{iL_{M_t}\Delta t/2} \cdots e^{iL_2\Delta t/2} e^{iL_1\Delta t} e^{iL_2\Delta t/2} \cdots e^{iL_{M_t}\Delta t/2}. \tag{4.4}$$

This formula is generalized to the formula with the M_t multiple-time steps as

$$\begin{aligned} e^{iL\Delta t} &= e^{iL_{M_t}(n_1 n_2 \cdots n_{M_t-1}\delta t/2)}[e^{iL_{M_t}(n_1 n_2 \cdots n_{M_t-2}\delta t/2)} \cdots \\ &\times [e^{iL_2(n_1\delta t/2)}[e^{iL_1\delta t}]^{n_1} e^{iL_2(n_1\delta t/2)}]^{n_2} \cdots \\ &\times e^{iL_{M_t}(n_1 n_2 \cdots n_{M_t-2}\delta t/2)}]^{n_{M_t-1}} e^{iL_{M_t}(n_1 n_2 \cdots n_{M_t-1}\delta t/2)} . \end{aligned} \tag{4.5}$$

The forces acted on a atom can be divided into five terms according to its periods of the fluctuation with time.

$$\begin{aligned} &\vec{F}(\vec{r}) = \vec{F}^{(1)}(\vec{r}) + \vec{F}^{(2)}(\vec{r}) + \vec{F}^{(3)}(\vec{r}) + \vec{F}^{(4)}(\vec{r}) + \vec{F}^{(5)}(\vec{r}) \\ &\vec{F}^{(1)}(\vec{r}) = \vec{F}_{stretch}(\vec{r}) \\ &\vec{F}^{(2)}(\vec{r}) = \vec{F}_{bend}(\vec{r}) + \vec{F}_{torsion}(\vec{r}) + \vec{F}_{Hbond}(\vec{r}) \\ &\vec{F}^{(3)}(\vec{r}) = \vec{F}_{vdW}^{near}(\vec{r}) + \vec{F}_{elec}^{near}(\vec{r}) \\ &\vec{F}^{(4)}(\vec{r}) = \vec{F}_{vdW}^{med}(\vec{r}) + \vec{F}_{elec}^{med}(\vec{r}) \\ &\vec{F}^{(5)}(\vec{r}) = \vec{F}_{vdW}^{far}(\vec{r}) + \vec{F}_{elec}^{far}(\vec{r}) . \end{aligned} \tag{4.6}$$

The near forces: $\vec{F}_{vdW}^{near}$, $\vec{F}_{elec}^{near}$ correspond to the direct computation pair-by-pair in Sect.3. The medium forces : $\vec{F}_{vdW}^{med}$, $\vec{F}_{elec}^{med}$ correspond to the far fields by the multipole expansions in upper level of the deepest level. The far forces : $\vec{F}_{vdW}^{far}$, $\vec{F}_{elec}^{far}$ correspond to the whole far fields except the far fields in the medium level.

The Liouville operators can be divided into five terms according to the periods of the variation of forces with time. Also, iL_1 can be divided into three terms.

$$iL = iL_{NHCP} + iL_{1a} + iL_{1b} + iL_2 + iL_3 + iL_4 + iL_5 \tag{4.7}$$

$$\begin{aligned} iL_s &= \sum_{i=1}^{N} \frac{\vec{F}_i^{(s)}(\vec{r})}{m_i} \cdot \vec{\nabla}_{\vec{v}_i} + \sum_{\alpha=1}^{3}\sum_{\beta=1}^{3} (\vec{G}_h^{(s)})_{\alpha\beta} \frac{\partial}{\partial(\vec{v}_h)_{\alpha\beta}} \qquad s = 2\text{-}5 \\ iL_{NHCP} &= -\sum_{i=1}^{N} v_{\xi_1}\vec{v}_i \cdot \vec{\nabla}_{\vec{v}_i} + \sum_{i=1}^{M} v_{\xi_i}\frac{\partial}{\partial \xi_i} + \sum_{i=1}^{M-1}(G_i - v_{\xi_i}v_{\xi_{i+1}})\frac{\partial}{\partial v_{\xi_i}} + G_M \frac{\partial}{\partial v_{\xi_M}} \\ &- \sum_{i=1}^{N}\left[\left(\vec{v}_h + \frac{\mathrm{Tr}[\vec{v}_h]}{3N}\vec{\mathrm{I}}\right)\vec{v}_i\right]\cdot\vec{\nabla}_{\vec{v}_i} + \sum_{\alpha=1}^{3}\sum_{\beta=1}^{3}[(\vec{G}_h^{(1)})_{\alpha\beta} - (\vec{v}_h)_{\alpha\beta}v_{\xi_1}]\frac{\partial}{\partial(\vec{v}_h)_{\alpha\beta}} \end{aligned} \tag{4.8}$$

$$\begin{aligned} iL_{1a} &= \sum_{i=1}^{N} \frac{\vec{F}_i^{(1)}(\vec{r})}{m_i} \cdot \vec{\nabla}_{\vec{v}_i} \\ iL_{1b} &= \sum_{\alpha=1}^{3}\sum_{\beta=1}^{3}(\vec{v}_h\vec{h})_{\alpha\beta}\frac{\partial}{\partial(\vec{h})_{\alpha\beta}} + \sum_{i=1}^{N}[\vec{v}_i + (\vec{v}_h\vec{r}_i)]\cdot\vec{\nabla}_{\vec{r}_i} \end{aligned} \tag{4.9}$$

where $\vec{G}_h^{(s)}$ is given by (4.10).

$$\begin{aligned} (\vec{G}_h^{(1)})_{\alpha\beta} &= \frac{1}{W}\sum_{i=1}^{N} m_i(\vec{v}_i)_\alpha(\vec{v}_i)_\beta \\ &+ \frac{1}{W}\left[\left(\frac{1}{3N}\sum_{i=1}^{N} m_i\vec{v}_i^2 - P_{ext}V\right)\delta_{\alpha\beta} + \sum_{i=1}^{N}(\vec{F}_i^{(1)})_\alpha(\vec{r}_i)_\beta - (\vec{\Phi}'^{(1)}\vec{h}')_{\alpha\beta}\right] \\ (\vec{G}_h^{(s)})_{\alpha\beta} &= \frac{1}{W}\left[\sum_{i=1}^{N}(\vec{F}_i^{(s)})_\alpha(\vec{r}_i)_\beta - (\vec{\Phi}'^{(s)}\vec{h}')_{\alpha\beta}\right] \qquad s = 2\text{-}5 \end{aligned} \tag{4.10}$$

The time evolution operation for the NTP ensemble is expressed by the following equation.

$$e^{iL\Delta t} = e^{i(n_1 n_2 n_3 n_4 \delta t/2)L_5}[e^{i(n_1 n_2 n_3 \delta t/2)L_4}[e^{i(n_1 n_2 \delta t/2)L_3}[e^{i(n_1 \delta t/2)L_2}$$
$$\times[e^{i(\delta t/2)L_{NHCP}} e^{i(\delta t/2)L_{1a}} e^{i\delta t L_{1b}} e^{i(\delta t/2)L_{1a}} e^{i(\delta t/2)L_{NHCP}}]^{n_1} \tag{4.11}$$
$$\times e^{i(n_1\delta t/2)L_2}]^{n_2} e^{i(n_1 n_2 \delta t/2)L_3}]^{n_3} e^{i(n_1 n_2 n_3 \delta t/2)L_4}]^{n_4} e^{i(n_1 n_2 n_3 n_4 \delta t/2)L_5}.$$

The multidimensional vector of independent variables Gfor the NTP ensemble system are expressed by (4.12).

$$\Gamma = \{\vec{r}_i, \vec{v}_i, \vec{h}, \vec{v}_h, \xi_j, v_{\xi_j}\} \tag{4.12}$$

When the operator : $e^{i(\Delta t_s/2)L_s}$ (s =2-5) in the right hand side of the formula (4.11) acts upon G(0), the update formula for velocity (4.13) is obtained. This update operation is repeated n_s times.

$$e^{i(\Delta t_s/2)L_s}\Gamma(0) = \exp\left[\frac{\Delta t_s}{2}\left(\sum_{i=1}^{N}\frac{\vec{F}_i^{(s)}}{m_i}\cdot\nabla_{\vec{v}_i} + \sum_{\alpha=1}^{3}\sum_{\beta=1}^{3}(\vec{G}_h^{(s)})_{\alpha\beta}\frac{\partial}{\partial(\vec{v}_h)_{\alpha\beta}}\right)\right]\Gamma(0)$$
$$= \{\vec{r}_i(0), \vec{v}_i(0) + \frac{\Delta t_s}{2}\frac{\vec{F}_i^{(s)}(0)}{m_i}, \vec{h}(0), \vec{v}_h(0) + \frac{\Delta t_s}{2}\vec{G}_h^{(s)}(0), \xi_j(0), v_{\xi_j}(0)\} \tag{4.13}$$
$$= \{\vec{r}_i(0), \vec{v}_i(\frac{\Delta t_s}{2}), \vec{h}(0), \vec{v}_h(\frac{\Delta t_s}{2}), \xi_j(0), v_{\xi_j}(0)\}.$$

Next, the operator : $e^{i(\delta t/2)L_{NHCP}}$ in the right hand side of (4.11) acts upon G(0), the update equations for $\vec{v}_i, \vec{h}, \vec{v}_h, \xi_j, v_{\xi_j}$ are obtained as follows [6]:

$$v_{\xi_m}\left(\frac{\delta t}{4}\right) = v_{\xi_m}(0)\exp\left(-\frac{\delta t}{4}v_{\xi_{m+1}}\right) + \frac{\delta t}{4}G_m\exp\left(-\frac{\delta t}{8}v_{\xi_{m+1}}\right)$$
$$\xi_m\left(\frac{\delta t}{2}\right) = \xi_m(0) + \frac{\delta t}{2}v_{\xi_m}\left(\frac{\delta t}{4}\right)$$
$$\vec{v}_h\left(\frac{\delta t}{4}\right) = \vec{v}_h(0)\exp\left(-\frac{\delta t}{4}v_{\xi_1}\right) + \frac{\delta t}{4}\vec{G}_h^{(1)}\exp\left(-\frac{\delta t}{8}v_{\xi_1}\right)$$
$$\vec{v}_i\left(\frac{\delta t}{2}\right) = \vec{C}_h^t \vec{I}_e \vec{C}_h \vec{v}_i(0) \tag{4.14}$$
$$\vec{I}_e = \exp\left(-\lambda_\alpha\frac{\delta t}{2}\right)\delta_{\alpha\beta}$$

where $\vec{C}_h, \vec{\lambda}$ are eigenvectors and eigenvalues of eigenequation (4.15).

$$\left(\vec{v}_h + \frac{\mathrm{Tr}[\vec{v}_h]}{3N}\vec{\mathrm{I}} + v_{\xi_1}\vec{\mathrm{I}}\right)\vec{C}_h = \vec{\lambda}\vec{C}_h \tag{4.15}$$

The operator : $e^{i(\delta t/2)L_{1a}}$ in the right hand side of the formula (4.11) acts upon G(0), the update equations for $\vec{v}_i$ are obtained as follows :

$$\vec{v}_i\left(\frac{\delta t}{2}\right) = \vec{v}_i(0) + \frac{\delta t}{2}\frac{\vec{F}_i^{(1)}(0)}{m_i}. \tag{4.16}$$

Finally, the operator : $e^{i(\delta t/2)L_{1b}}$ in the right hand side of (4.11) acts upon G(0), the update equations for $\vec{r}_i, \vec{h}$ are obtained as follows :

$$\vec{r}_i(\delta t) = \vec{C}_h^t\{\vec{I}_e\vec{C}_h\vec{r}_i(0) + \delta t\,\vec{I}_s\vec{C}_h\,\vec{v}_i(\delta t/2)\}$$
$$\vec{h}(\delta t) = \vec{C}_h^t\vec{I}_e\vec{C}_h\vec{h}(0) \qquad (4.17)$$
$$\vec{I}_e = \exp\left(\lambda_\alpha \frac{\delta t}{2}\right)\delta_{\alpha\beta}$$
$$\vec{I}_s = \exp\left(\lambda_\alpha \frac{\delta t}{2}\right)\delta_{\alpha\beta}\frac{\sinh(\lambda_\alpha \delta t/2)}{(\lambda_\alpha \delta t/2)}$$

where $\vec{C}_h, \vec{\lambda}$ are eigenvectors and eigenvalues of eigenequation: $\vec{v}_h\,\vec{C}_h = \vec{\lambda}\vec{C}_h$.

5. Applications

In this article, we have applied a new accelerated molecular dynamics method using the reversible multiple time step integrators and fast multipole expansion method to an antiferroelectric liquid crystalline molecule (MHPOBC) to understand the origin of bent structure formation and the molecular packing property of MHPOBC in a crystalline phase. The MHPOBC system is composed of 12240 atoms which are arranged in the MD cell with the sizes of 40 Å× 30 Å×83 Å shown in Fig. 2. We have used the two levels fast multipole expansion method with the square cell of 10 Å×10 Å×10 Å in the deepest level. We have adapted a time step size of 0.25fs for the bond and bond angle forces, a time step size of 2.0 fs for the torsion force and the direct computation part of FMM, the time step size of 8.0 fs for the multipole part of FMM. We obtained an acceleration of about 46 times over the conventional method. Elsewhere we will describe the results of the MHPOBC system [13] and the efficiency of a new accelerated molecular dynamics method [14] in detail.

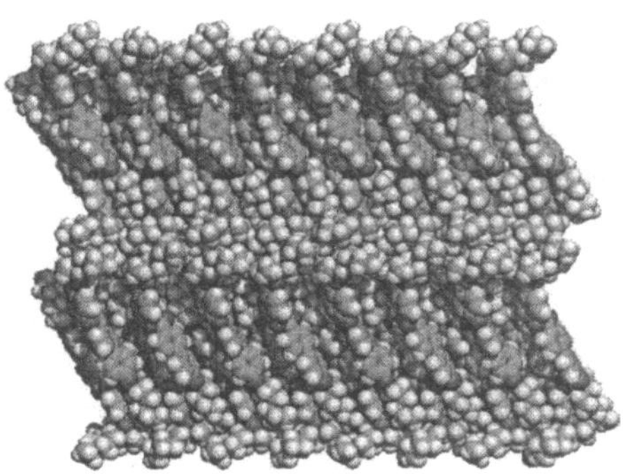

Fig. 2. The crystal structure of MHPOBC obtained by molecular dynamics simulation at 300 K and 0.1 MPa.

6. Summary

A new accelerated molecular dynamics method for the NTP ensemble system with up to a million atoms and molecules has been derived using the combination of the fast multipole expansion method and the reversible multiple time step integrators. We have applied this new accelerated molecular dynamics method to an antiferroelectric liquid crystal, MHPOBC, which is composed of 12240 atoms, and obtained an acceleration of about 46 times ovwe the conventional method.

References

1. L. Greengard and V. I. Rokhlin: J. Comp. Phys. **73**, 325 (1987).
2. K. E. Schmidt and M. A. Lee: J. Stat. Phys. **63**, 1223 (1991).
3. H. Ding, N. Karasawa, and W. A. Goddard III: J. Chem. Phys. **97**, 4309 (1992).
4. M. Tuckerman, B. J. Berne, and G. J. Martyna: J. Chem. Phys. **97**, 1990 (1992).
5. P. Procacci and B. J. Berne: Mol. Phys. **83**, 255 (1994).
6. G. J. Martyna, M. Tuckerman, D. J. Tobias, and M. L. Klein: Mol. Phys. **87**, 1117 (1996).
7. H. C. Andersen: J. Chem. Phys. **72**, 2384 (1980).
8. M. Parrinello and A. Rahman: J. Appl. Phys. **52**, 182 (1981).
9. S. Nose: J. Chem. Phys. **81**, 511 (1984).
10. W. H. Hoover: Phys. Rev. A **31**, 1695 (1985).
11. G. J. Martyna, M. L. Klein, and M. Tuckerman: J. Chem. Phys. **97**, 2635 (1992).
12. G. J. Martyna, D. J. Tobias, and M. L. Klein, J. Chem. Phys. **101**, 4177 (1994).
13. Mikami: J. Phys. Chem. (to be published).
14. Mikami: J. Comp. Chem. (to be published).

Program Tuning for Large-Scale Simulations in Computational Materials Science

Hiroshi Mizuseki, Ryoji Sahara, Zhi-Qiang Li, Kaoru Ohno, and Yoshiyuki Kawazoe

Institute for Materials Research, Tohoku University, Sendai 980-8577, Japan

Summary. In this paper, several examples of program tuning for vector and parallel machines and their efficiencies are studied. For the vector machine, we modify a Monte Carlo code to determine its practical maximum speed, and we obtain 100 times faster calculation than the RISC based machine. Furthermore, we try to parallelize an *ab initio* molecular dynamics calculation code, and measure the efficiency of a parallel machine.

Key Words. Vectorization, parallelization, Monte Carlo simulation, *ab initio* molecular dynamics, vector machine, parallel machine

1. Introduction

The main purpose of our simulation group is to establish an environment to predict properties of new materials before experimental measurement. Towards this aim, we are developing computer codes to simulate electronic, magnetic and mechanical properties of new materials at an atomistic level. Moreover, the statistics and dynamics of materials under various conditions are studied. For these purposes, the highest speed machine in the world should be employed. Recently, a variety of large-scale simulations were carried out using a vector supercomputer HITAC S-3800/380. In these studies, we vectorize various molecular dynamics codes, and we developed a complex Monte Carlo (MC) code for the vector machine. However, a few years ago the developmental activities of conventional vector machines almost stopped, because of large energy usage and high developmental cost. Moreover, the number of processors contained in a vector machine is usually limited to 32, for example in a CRAY T90. The main trend of computer technology is moving toward parallel machines with a large number of RISC architecture processor elements. Recently, a new vectorization feature became available (pseudo-vector processing function) on a Hitachi parallel machine, although its architecture is RISC. Conventional RISC processors run into difficulty when data for large-scale numerical calculations can not fit in the cache. This facility, the so-called "pseudo-vector processing function" is a pipelining technique which fetches data directly from main memory to floating point registers without holding up the execution. This facility can significantly improve performance for large-scale numerical calculations. As a result, in the near future, the main platform of scientific simulations is expected to be parallel machines. Simultaneously, the scale of our simulations has gone beyond the ability of conventional vector machines and the memory size of one RISC chip. It is necessary that we prepare programs that are well tuned for the parallel machine and investigate the processing speeds of the two different kinds of machines: vector and parallel machines.

The supercomputer we are using for materials science is HITAC S-3800/380 vector machine [1] with 3 tightly connected CPUs; each CPU has 8 GFLOPS processing speed and 2 GB common main memory. It has a further 14 GB of extended memory.

Our parallel machine is a HITAC SR2201 [2]. This machine has the previously described pseudo-vector processing function and a 3-dimensional crossbar network. These specifically developed features support high-speed numerical calculations with large scale memory usage distributed on multiple processing elements and achieve an excellent cost/performance ratio.

The main aim of this paper is to show a few examples of how to use different machines effectively and how to prepare for a next generation supercomputer. The present paper is organized as follows: Calculation models and details of tuning are introduced in Sect. 2. In this section, we develop the MC code for the test of vectorization and *ab initio* MD for the test of parallelization. A large computer simulation is performed and the calculation speeds of vector and parallel machines are given in Sect. 3.

2. Calculation Schemes

In a previous paper [3], we proposed a bcc lattice model, which enables us to describe the melting process of Si. In this study, we observed a phase transition and Ostwald growth by performing large scale and long MC steps. In this paper, we report the efficiency of vectorization of this MC code.

We chose *ab initio* molecular dynamics for a calculation scheme to test the efficiency of parallelization. This calculation is performed on 1 processor of the supercomputer and up to 64 processors of the parallel computer.

Detailed descriptions of the calculation scheme and scientific results are provided elsewhere: Monte Carlo method in [3] and *ab initio* molecular dynamics in [5].

2.1 Monte Carlo Method

The present lattice model is based on the assumption that each atom is allowed to move only on the bcc lattice sites. A 50% filling on the bcc lattice leads to the diamond structure for the ordered phase. As an example, 2×2×2 bcc cells are shown in Fig. 2.1. Figure 2.1a shows a sample of a liquid state, where atoms occupy bcc sites in disorder. In contrast, Fig. 2.1b is the case where the atoms occupy these sites in order and the diamond structure is realized. The energy per atom in this model is defined by nearest neighbor configurations.

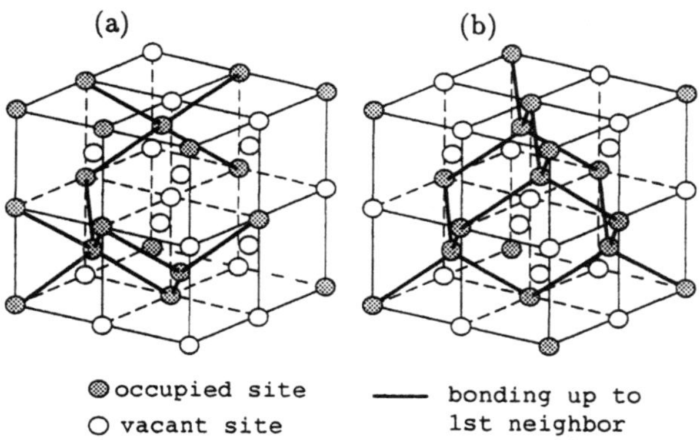

Fig. 2.1. Schematic diagram of bcc lattice model with 50% filling.

Each atom moves one by one to the nearest neighbors, and the model is thus similar to the Ising model on the bcc lattice sites with Kawasaki spin exchange algorithms. Therefore, the i–th trial affects the i+1–th trial, and all movements of atoms are sequential. Figure 2.2 shows the flowchart of this "sequential" algorithm. One repeats this movement of atoms again and again by the following steps.

1. Select one lattice site i at which the atom is considered to move to the nearest site.

2. Compute the energy change δE associated with that move. To reduce the number of calculations, only nearest neighbor sites surrounding the nearest sites of the trial site are selected, because only this small area is affected by one trial. This part becomes the "inner loop" whose length is very short.

3. The normal Metropolis method is used to decide whether this trial is accepted or rejected. That is, draw a random number ξ uniformly distributed between 0 and 1, then, if $\xi < \exp(\delta E/k_B T)$ (k_B : Boltzmann constant) move the atom, otherwise do not move it.

In this algorithm, the trial operation can be branched. The algorithm is usually written in a form including the "inner loop" which does not vectorize effectively, because usually the number of nearest neighbors, which can be vectorized, is very small. This means that some reorganization of the algorithm is needed.

One technique of vectorizing on the MC algorithm is explained in [6]. The general idea is to try to decompose the whole lattice into the least number of sublattices, on each pair of which the given dynamics may be performed independently. To explain this algorithm, we consider a simple spin-flip Ising square lattice with nearest neighbor interactions, and then a spin-exchange system.

For the spin-flip algorithm, the entire system is divided into two interpenetrating sublattices in which each sublattice consists of sites which are part of a next-nearest-neighbor network, as shown in Fig. 2.3. This approach is sometimes called the "checkerboard algorithm", since no two spins on the same sublattice interact, the entire spin-flip operations are applied in a single vector operation for each sublattice.

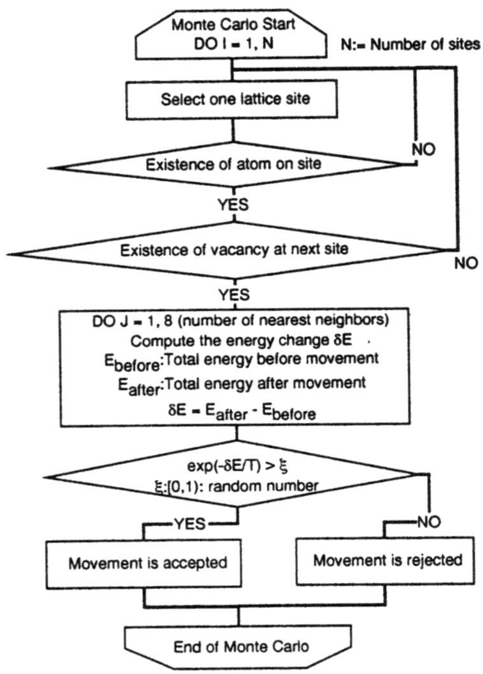

Fig. 2.2. Flowchart of non-vectorization version of bcc model. Main part is sequential procedure, because these energy calculations are affected by other positions. Unfortunately, this inner loop length is very short.

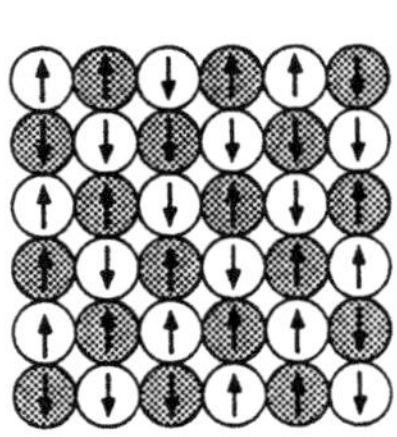

Fig. 2.3. Sublattices of Ising model.

The spin-exchange Kawasaki algorithm is vectorized by extending the "checkerboard algorithm". Each checkerboard sublattice is further divided into additional sublattices so as not to interact with each other on the same sparse sublattice. At first, the simple Ising square lattice with nearest neighbor interactions is introduced to explain the idea of the extended "checkerboard algorithm". Here, keep in mind that on-site energy is defined by configurations composed of surrounding nearest neighbors. We must consider how many sublattices are affected when one atom moves to a nearest neighbor site.

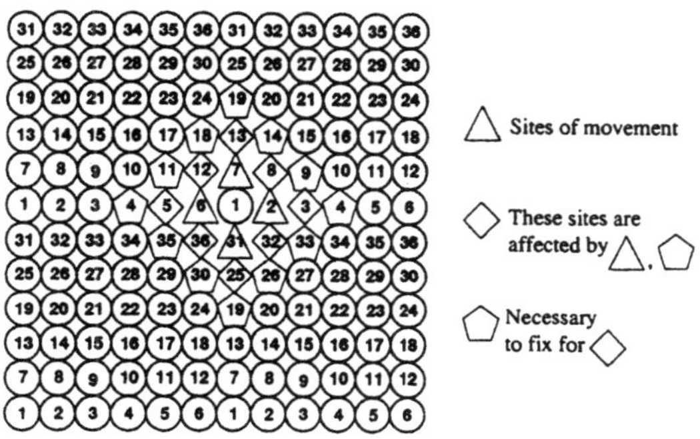

Fig. 2.4. Sublattices of the Kawasaki model.

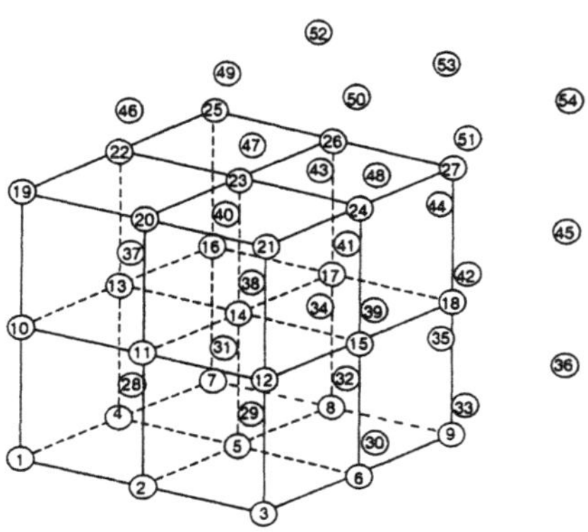

Fig. 2.5. 54 sublattices for vectorization in bcc model. The movement on each lattice does not affect the energy of others.

The answer is obtained by considering how many on-site energies are changed. Figure 2.4 shows the sublattices which are needed to satisfy this condition. For example, one spin on sublattice 1 is allowed to move to one of the nearest neighbor sites, sublattice 2, 6, 7, or 31. On-site energy of these triangular nearest neighbor sites in the triangle are defined by these neighbors (diamond and circle). Furthermore, the on-site energies of these sites are defined by these nearest neighbors, some of which are shown by pentagons, whose energy must not be changed by an exchanging atom on sublattice 1. So as not to affect the spin exchange for "on-site energy" on the same sublattice, at least 6 intervals are necessary for vertical and horizontal directions, respectively. Now, we know that $6 \times 6 = 36$ sublattices are needed for satisfying the condition. Applying this idea, a small length inner loop can be extended to the outer loop and the code is effectively vectorized.

We applied the algorithmic optimization to the present bcc lattice model. The bcc lattice model with nearest neighbor interactions must be divided into 54 sublattices constructing 3×3×3 bcc unit cells. (see Fig. 2.5.) Figure 2.6 shows a flowchart of the vectorized model. The inner loop length is the total number of sites N divided by the number of sublattices, 54.

Note that even if a model contains more distant neighbor interactions, the method may still be implemented by subdivision of the lattice into more sublattices, however, as the number of sublattices grows, the number of sites per sublattice decreases for any constant lattice size and hence the length of the vector loop becomes smaller.

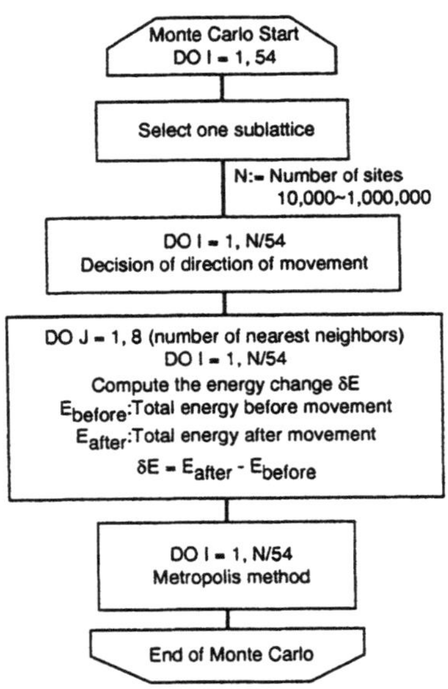

Fig. 2.6. Flowchart of vectorization version of bcc model. Under the main operation, the energy calculation is independent of other atoms since each movement does not affect other movements. Therefore, the program code is vectorized. In a large-scale system, we can obtain a long loop length. This loop length is N/54.

2.2 *Ab initio* Molecular Dynamics

We use a first-principle code based on a plane wave expansion and a norm-conserving pseudopotential, to find an optimal adsorption height of one C_{60} molecule adsorbed on the valley of the GaAs 2×4 surface. In a tetragonal unit cell, we put 40 As and 44 Ga atoms (altogether 6 layers), fixing all the positions at and near the surface.

3. Results

3.1 Monte Carlo Method

Figure 3.1 shows the increase in calculation speed by vectorization as a function of memory size, which is proportional to the number of atoms. The efficiency is normalized by a variable number of atoms per second. The closed circles are the result of a vectorized program running on a single processor in HITAC S-3800. The open circles are the non-vectorized program (this code is suitable for the RISC chip machine) on S-3800. The open square are the non-vectorized program running on the DEC7000. The DEC7000 is a conventional RISC chip machine with 2 MB 2nd cache. When the memory size used is less than 2 MB (corresponding to only a 16000 atom system), the vector machine and the RISC based machine produce almost the same calculation speed. When the memory size exceeds 2 MB, cache miss-hit occurs on the RISC chip machine and the speed is reduced, although the speed on the vector machine does not change with the memory size within these simulation sizes. By using a vector tuning code on the S-3800, a peak speed of 650000 spin-exchange trials per second was achieved. Since we use different random number generators in the two programs, it is clear that we can not obtain exactly the same results, even if the same conditions are considered. However, the statistical averages of the results are the same within the statistical error.

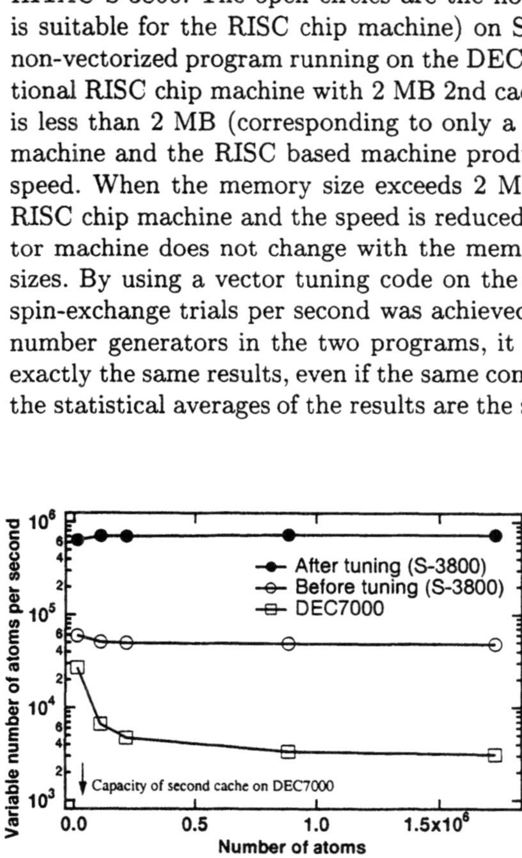

Fig. 3.1. Effect of vectorization for vector machine.

Now, we turn back to the simulation results. The system size we usually studied is 16×16×16 bcc lattice sites (4096 atoms). It is assumed that only the atoms which construct the tetrahedral structure are lower in energy (we call this assumption the Tetrahedron model). At low enough temperatures, the potential model yields the diamond structure at thermal equilibrium after starting from the liquid state. Figure 3.2 shows the final distributions of n-fold coordinated atoms as a function of temperature after 20000 MC steps.

Figure 3.2a is the result starting from the liquid state and Fig. 3.2b from the solid state. It is confirmed that the 4–fold coordinated atoms (diamond structure) mainly exist in the low temperature region. Suddenly, at around a reduced temperature of 6 in Fig. 3.2a and around 7 in Fig. 3.2b, higher coordinated atoms appear instead of the 4–fold ones. By taking account of the difference of the initial conditions, the 1st-order phase transition involving a super-cooled state with a hysteresis is represented. However, hysteresis behavior is shown quantitatively, a finite size efficiency must be considered to be qualitative. Larger size simulation is necessary to consider them. Such a study is in progress in our group, and will be published elsewhere [4].

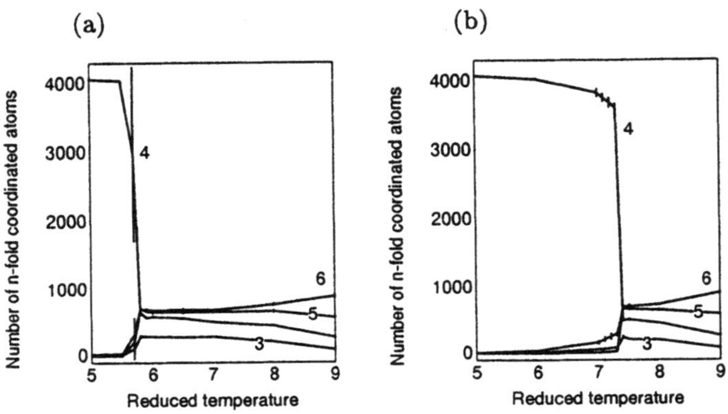

Fig. 3.2. Final distribution of n–fold atoms as a function of reduced temperature after 20000 MC steps.

3.2 *Ab initio* Molecular Dynamics

By *ab initio* molecular dynamics, we investigated adsorption of C_{60} on a GaAs(001) 2×4 surface [5]. From this calculation, we obtain the isodensity surface of the resulting total charge distribution. This result is important to explain the dimerized C_{60} film experimentally grown on a GaAs(001) surface up to 10 layers [7], which is observed very specifically only on a GaAs surface. From this distribution, we can estimate how many electrons are transferred from As atoms located on the surface to the first layer of C_{60}s. The estimated amount of the charge transfer is about 1.76 electrons per C_{60}.

Figure 3.3 shows the performance of a parallelized program for the first principle calculation. The achieved performance of 2.97 GFLOPS with 64 processors is almost the same as the most powerful 1 CPU of the vector supercomputer of 8 GFLOPS. Figure 3.4 shows the ratio of communication time to the total calculation time.

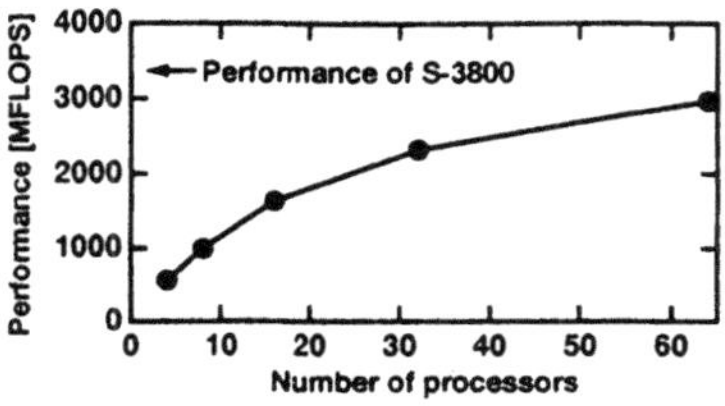

Fig. 3.3. Parallel efficiency of first principle calculation as a function of number of PE.

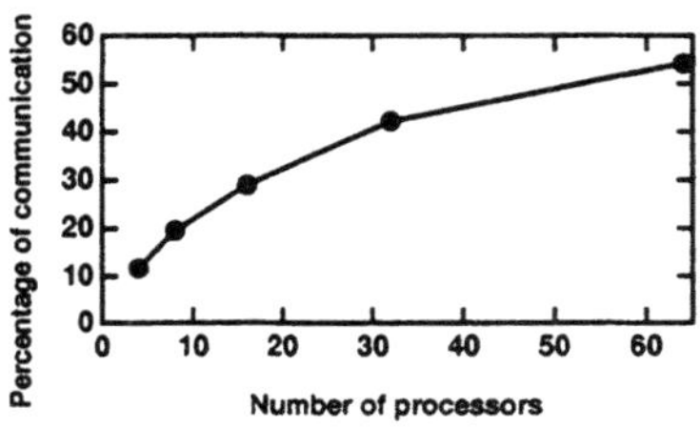

Fig. 3.4. Percentage of communication time in total CPU time as a function of number of PE.

4. Discussion

For computer simulations in the field of materials science, supercomputers will never be held to be sufficiently fast since the number of atoms which should be treated is on the order of 10^{24}. From a pragmatic viewpoint, we should pursue maximum utilization of existing supercomputers. However, since computer technology is evolving daily, we should construct new methods of simulations which are suitable for those innovative computer architectures. For example, the tight-binding method can be parallelized easily in a rather straightforward way. On the other hand, diagonalization of a large matrix, which is the most time consuming part of the *ab initio* quantum method, is known to be difficult to be parallelized. To change the basic formulation of the problem is the most probable way to accomplish parallelization.

The theoretical maximum performance of RISC architecture can be obtained only when an entire operand array is stored in their cache. On the other hand, vector processors can transfer all operands needed to and from CPUs without the disadvantage of cache-miss. Noting these issues, the band-width between registers and local main memory as well as the band-width of interprocessor communications should be fundamentally enlarged. The pseudo-vector processing function employed in SR2201 provides a significant reduction of cache-miss of large arrays and improves the ability of data transfer compared to standard RISC systems, and most other parallel machines.

5. Conclusions

Large scale computer simulations in materials science have been performed on a vector supercomputer and on a parallel computer. We vectorize the MC code to obtain the advantages of a vector machine. As a result, we can calculate 100 times faster than a RISC based machine. For the first principle calculation, because the main part of the heavy routine is matrix diagonalization, it is not easy to parallelize the existing code. We are developing a new formulation suitable for a parallel machine. Measured efficiencies are satisfactory up to 64 processors.

Acknowledgements

The authors would like to express their sincere thanks to the Materials Information Science Group of the Institute for Materials Research, Tohoku University, for their continuous support of the HITAC S-3800/380 and SR2201 supercomputing facilities. They are also grateful to the Hitachi Company for supporting parallelization of our FORTRAN source codes.

References

1. K. Ishii, H. Abe, S. Kawabe, M. Hirai: Informatik aktuell H.-W. Meuer (Hrsg.) SUPERCOMPUTER '92, Anwendungen, Architekturen, Trends Seminar, Mannheim, 25-27. June 1992 (Springer-Verlag), pp. 65–81 "An Overview of the HITACHI S-3800 Series Supercomputer".
2. K. Nakazawa, H. Nakamura, H. Imori, S. Kawabe: Proc. Supercomputing '92 (IEEE/ACM), pp.642–651 (Nov. 1992); H. Nakamura, H. Imori, K. Nakazawa, T. Boku, I. Nakata, Y. Yamashita, H. Wada, Y. Inagami: Proc. Int. Conf. on Supercomputing '93 (ACM), pp.298–307 (1993).
3. R. Sahara, H. Mizuseki, K. Ohno, S. Uda, T. Fukuda, Y. Kawazoe: Sci. Rep. RITU **A43**, 23–28 (1997).
4. R. Sahara, H. Mizuseki, K. Ohno, S. Uda, T. Fukuda, Y. Kawazoe: (to be submitted).
5. K. Ohno, Z. Q. Li, H. Kamiyama, Y. Kawazoe, Q. Xue, T. Hasegawa, H. Shinohara, T. Sakurai: Sci. Rep. RITU **A43**, 61–65 (1997).
6. K. Binder (ed.), *Topics in Applied Physics* Vol. 71, *The Monte Carlo Simulation in Condensed Matter Physics*, (Springer-Verlag, 1992), pp. 22–51, *Vectorisation of Monte Carlo Programs for Lattice Models Using Supercomputers*, by D. P. Landau.
7. Q. Xue, T. Hashizume, Y. Hasegawa, H. Kamiyama, Z.-Q. Li, K. Ohno, Y. Kawazoe, H. Shinohara, T. Sakurai: (to be submitted).

Subject Index

MIX
Papier aus verantwortungsvollen Quellen
Paper from responsible sources
FSC® C105338

If you have any concerns about our products,
you can contact us on
ProductSafety@springernature.com

In case Publisher is established outside the EU,
the EU authorized representative is:
Springer Nature Customer Service Center GmbH
Europaplatz 3, 69115 Heidelberg, Germany

Printed by Libri Plureos GmbH
in Hamburg, Germany